# WONDERS OF LIFE

For George Albert Eagle, a Wonder of Life

—Brian Cox

To my beautiful Anna, it's a precious thing, even more so when you share it.
Thank you for our first ten years and for all your endless love, support and guidance.

—Andrew Cohen

First published in the United Kingdom by Collins, an imprint of
HarperCollinsPublishers, by arrangement with the BBC.

Text © Brian Cox and Andrew Cohen 2013

Photographs (with the exception of those detailed on page 287) © BBC

Infographics, diagrams, design, and layout © HarperCollins Publishers 2013

The BBC logo is a trademark of the British Broadcasting Corporation
and is used under licence.

BBC logo © BBC 1996

HarperCollins books may be purchased for educational, business, or sales
promotional use. For information please write: Special Markets Department,
HarperCollinsPublishers, 10 East 53rd Street, New York, NY 10022.

ISBN 978-0-06-223883-2

Printed by South China Printing Company in China

Distributed throughout the world by
HarperCollinsPublishers
10 East 53rd Street
New York, NY 10022
Fax: (212) 207-7654

# WONDERS OF LIFE

## EXPLORING THE MOST EXTRAORDINARY
## FORCE IN THE UNIVERSE

# BRIAN COX

### AND ANDREW COHEN

HARPER DESIGN

An Imprint of HarperCollins Publishers

INTRODUCTION
—
# WONDERS OF LIFE
—

CHAPTER 1
—
# HOME
—

CHAPTER 2
—
# WHAT IS LIFE?
—

# INTRODUCTION

# WONDERS OF LIFE

The Nobel Prize-winning physicist Richard Feynman used to tell a story about an artist friend who challenged him about the beauty of a flower. 'You as a scientist, oh, take this all apart and it becomes a dull thing' he said. Feynman, after describing his friend as 'kind of nutty', went on to explain that whilst the aesthetic beauty of nature is surely open to everyone, albeit not in quite as refined a way, the world becomes more beautiful as our understanding deepens.

The flower is made up of cells, single units with identical genes. Hidden within are a multitude of biochemical machines, each highly specialised to perform complex tasks that keep the cell alive. Some contain chloroplasts, once free-living bacteria, co-opted into capturing light from the Sun and using it to assemble food from carbon dioxide and water. There are mitochondria, factories that pump protons up energy 'waterfalls' and insert organic waterwheels into the ensuing cascade to assemble ATP molecules – the universal batteries of life. And there is DNA, a molecule with a code embedded in its structure that carries the instructions to assemble the flower, but also contains fragments of the story of the origin and evolution of all life on Earth, from its beginnings 3.8 billion years ago to the endless forms most beautiful that have transformed a once-sterile world into the grandest possible expression of the laws of nature. This is beauty way beyond the aesthetic that, as Feynman concluded, 'only adds to the excitement and mystery and the awe of a flower. It only adds; I don't understand how it subtracts.'

I confess that, when we began thinking about filming *Wonders of Life*, my knowledge of biology was a little dated – I gave it up as an academic subject in 1984. As I recall, the idea for the series came from an off-hand reference I made to Andrew Cohen about a little book I had read as a physics undergraduate.

*What is Life?* is an account of a series of lectures given by the physicist Erwin Schrödinger, published in 1944. Schrödinger was a Nobel Prize winner, one of the founders of quantum theory, and a deep and high-precision thinker. In the book, he poses a simple yet profound question: 'How can the events in space and time which take place within the spatial boundary of a living organism be accounted for by physics and chemistry?' This question is beautifully phrased. Most important is the word 'How' at the beginning. Without this word, the question is metaphysical, in the sense that the answer may be 'No' – a complete understanding of life

may be forever beyond the natural sciences because there is something inherently supernatural about it. The word 'How' transforms it, and provides a significant and important insight into the mind of a scientist. Let us find out, by studying nature, developing theories and testing those theories against our observations of the living world, how life can be fully explained by the laws of physics and chemistry, as it surely must be. This, I submit, is an excellent description of the science of biology.

*Wonders of Life* might be best described as a series exploring our current understanding of Schrödinger's 'How' question. I enjoyed making the films immensely, because virtually everything in them was discovered after I gave up biology in 1984. The rate of discovery, driven by powerful new experimental techniques such as the exponentially increasing ease and decreasing cost of DNA sequencing, is quite dazzling and, Higgs Boson notwithstanding, I might be convinced that the 21st century has already become the century of the Life Sciences; but only 'might'.

A truly wonderful exception to the modernity is Darwin's theory of evolution by natural selection, published in November 1859 and spectacularly verified as a conceptual framework to understand the diversity and complexity of life on Earth. To understand Darwin's genius, look out of your window at the living world. Unless you are in the high Atacama Desert, you will surely see a living world of tremendous complexity. Even a blade of grass should be seen through Feynman's reductionist prism as a magnificent structure. On its own, it is a wonder, but viewed in isolation its complexity and very existence is inexplicable. Darwin's genius was to see that the existence of something as magnificent as a blade of grass can be understood, but only in the context of its interaction with other living things and, crucially, its evolutionary history. A physicist might say it is a four-dimensional structure, with both spatial and temporal extent, and it is simply impossible to comprehend the existence of such a structure in a universe governed by the simple laws of physics if its history is ignored.

And whilst you are contemplating the humble majesty of a blade of grass, with a spatial extent of a few centimetres but stretching back in the temporal direction for almost a third of the age of the Universe, pause for a moment to consider the viewer, because what is true for the blade of grass is also true

> *'How can the events in space and time which take place within the spatial boundary of a living organism be accounted for by physics and chemistry?'*
>
> *Erwin Schrödinger*

for you. You share the same basic biochemistry, all the way down to the details of proton waterfalls and ATP, and much of the same genetic history, carefully documented in your DNA. This is because you share a common ancestor. You are related. You were once the same.

I suppose this is a most difficult thing to accept. The human condition seems special; our conscious experience feels totally divorced from the mechanistic world of atoms and forces, and perhaps even from the 'lower forms' of life. If there is a central argument through the five films and chapters in *Wonders of Life*, it is that this feeling is an emergent illusion created by the sheer complexity of our arrangement of atoms. It must be, because the fundamental similarities between all living things outweigh the differences. If an alien biochemist had only two cells from Earth, one from a blade of grass and one from a human being, it would be immediately obvious that the cells come from the same planet, and are intimately related. If that sounds unbelievable, then this book is an attempt to convince you otherwise.

I write this in full appreciation of the so-called controversy surrounding Darwin's theory of evolution by natural selection. My original aim was to avoid the matter entirely, because I think there are no intellectually interesting issues raised in such a 'debate'. But during the filming of this series I developed a deep irritation with the intellectual vacuity of those who actively seek to deny the reality of evolution and the science of biology in general. So empty is such a position, in the face of evidence collected over centuries, that it can only be politically motivated; there is not a hint of reason in it. And more than that, taking such a position closes the mind to the most wonderful story, and this is a tragedy for those who choose it, or worse, are forced into it through deficient teaching.

As someone who thinks about religion very little – I reject the label atheist because defining me in terms of the things I don't believe would require an infinite list of nouns – I see no necessary contradiction between religion and science. By which I mean that if I were a deist, I would claim no better example of the skill and ingenuity of The Creator than in the laws of nature that allowed for the magnificent story of the origin and evolution of life on Earth, and their overwhelmingly beautiful expression in our tree of life. I am not a deist, philosopher or theologian, so I will make

no further comment on the origin of the laws of nature that permitted life to evolve. I simply don't know; perhaps someday we will find out. But be in no doubt that laws they are, and Darwin's theory of evolution by natural selection is as precise and well tested as Einstein's theories of relativity.

If this sounds a little strong, then perhaps it reveals my genuine excitement in learning about the sheer explanatory power of Darwin's theory when coupled with recent advances in biochemistry and genetics. Modern biology is close, in my view, to answering Schrödinger's 'How' question. There are unknowns to be sure, which is what makes the subject of these films doubly exciting. Some parts are speculative, but that is nothing to be ashamed of in science. Indeed, all science is provisional. When observations of nature contradict a theory, no matter how revered, ancient or popular, the theory will be unceremoniously and joyously ditched, and the search for a more accurate theory will be redoubled. The magnificent thing about Darwin's explanation of the origin of species is that it has survived over a hundred and fifty years of precision observations, and in that it has outlasted Newton's law of universal gravitation.

My favourite moment in the series is the final scene of the final film, which unusually, was filmed on our final evening; television shows are rarely made in chronological order. We found a tiny rocky island off the coast of northern Madagascar, no bigger than the average suburban garden, isolated in the warm waters of the Mozambique Channel. The idea was to sit down and chat about the experience of making the series, and film the result. I won't tell you what I thought and said, because that should wait until the end of the book. But I do want to say one thing here in the introduction. I recall a conversation in March 2009, just before we started filming *Wonders of the Solar System*. Andrew, my co-author and executive producer, said that we would have achieved our goal if those who watched never again looked at the night sky in quite the same way. This is in the spirit of Feynman's flower. Deeper understanding confers that most precious thing – wonder. A sky filled with tiny, twinkling lights is one thing, but a sky filled with other worlds is quite another. I have known this for virtually all my life, because I have always been an astronomer at heart. Perched on my island, thinking about what to say, I realised that I now felt precisely the same about a single blade of grass. ◉

# CHAPTER 1

## HOME

On Christmas Eve 1968, Frank Borman, Jim Lovell and William Anders became the first humans in history to lose sight of their home planet as they orbited the Moon on board Apollo 8. As Borman looked into the crystal dark, pitted by the faint light of a billion worlds untarnished by atmospheric gases, framed by a virgin lunar surface unseen since its formation 4.5 billion years ago, he described his universe without Earth as a 'vast, lonely, forbidding expanse of nothing'. On the ninth orbit, the crew made a scheduled live television broadcast in which they chose to read the Genesis creation story back across the quarter of a million miles to Earth.

**OPPOSITE:** 'Earth rise', first observed from Apollo 11 in 1969, gave us a totally new perspective on the planet we call home.

*'We are now approaching lunar sunrise and, for all the people back on Earth, the crew of Apollo 8 has a message that we would like to send to you:*
*In the beginning God created the heavens and the Earth.*
*And the Earth was without form, and void; and darkness was upon the face of the deep.'*

The act of reading from the Bible proved controversial, and was challenged in court as a violation of the 1st amendment of the United States Constitution, which prevents the promotion of religion by the federal government, of which NASA is a part. The Supreme Court dismissed the case, on the grounds that it had no jurisdiction in lunar orbit.

While the Genesis story is a myth, I have always found this broadcast moving; not merely because the King James version of the Bible contains some of the greatest prose ever written in the English language, but because it speaks to an ancient, resonant desire to understand our origin and the origin of our home. Why is the Earth a living oasis amid, as far as anyone can tell, a forbidding expanse of nothing? What is special about our pale blue anomaly of a world that makes it home to life?

These questions are complex, and we do not yet have all the answers, but there is a scientific consensus on at least some of the ingredients a planet requires to allow the emergence of life and the evolution of complex organisms capable of taking their first, faltering steps into a wider Universe. Many of those ingredients are common throughout the Solar System and beyond, but we have as yet no evidence for life, simple or complex, beyond Earth. That may be because the emergence of living things required a significant slice of luck and billions of years of relative stability; spacecraft builders may be a rare and precious commodity.

This thought may have been adrift somewhere in Frank Borman's consciousness, catalysed by his feelings of isolation 400,000 km from home, when he ended the 1968 Christmas broadcast with a phrase I have always found overpowering in its simplicity and depth of meaning. To me, it was an instinctive plea to all of us to value our home – the absolutely necessary platform for the continued existence of, just possibly, the only living civilisation in the Universe:

*'And, from the crew of Apollo 8, we close with good night, good luck, a Merry Christmas – and God bless all of you, all of you on the good Earth.'* ◉

# EVOLUTION OF LIFE

**SYNAPSID REPTILES**
Permian (299–251 MYA)
Large terrestrial 'mammal-like reptiles' become plentiful

**LAND VERTEBRATES**
Carboniferous (359–299 MYA)
First reptiles prosper due to the emergence of the amniote egg

**LOBE-FINNED FISH**
Devonian (416–359 MYA)
375 MYA Tiktaalik comes onto land, with features similar to those of (4-legged) tetrapods

**FERNS**
Devonian (416–359 MYA)
First clubmosses, horsetails, and ferns appear. The first seed-bearing plants, trees and wingless insects emerge

**JAWED FISH**
Silurian (443–416 MYA)
The jawed fish dominate the sea, and are the ancestors of all terrestrial vertebrates

**MILLIPEDES**
Silurian (443–416 MYA)
Thought to be among the first animals to have colonised the land, these arthropods probably ate fungi and detritus

**LONG STRAIGHT-SHELLED CEPHALOPODS**
Ordovician (488–443 MYA)
Larger than their Cambrian counterparts, most of which did not survive

**ANOMALOCARIDIDS**
Cambrian (542–488 MYA)
The largest Cambrian animals known, most of these marine animals were probably carnivorous

**TRILOBITES**
Cambrian (542–488 MYA)
Among the most successful of the earliest arthropods, roaming the oceans for over 270 million years

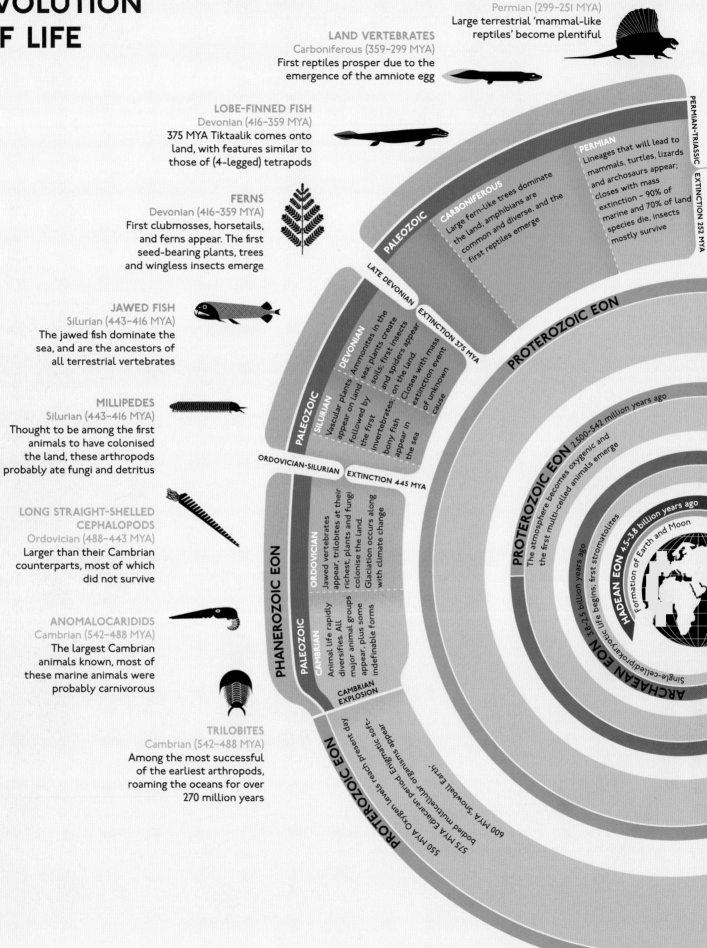

PERMIAN-TRIASSIC EXTINCTION 252 MYA

PERMIAN
Lineages that will lead to mammals, turtles, lizards and archosaurs appear; closes with mass extinction – 90% of marine and 70% of land species die, insects mostly survive

PALEOZOIC

CARBONIFEROUS
Large fern-like trees dominate the land; amphibians are common and diverse, and the first reptiles emerge

LATE DEVONIAN EXTINCTION 375 MYA

PALEOZOIC

DEVONIAN
Vascular plants appear on land followed by the first invertebrates; bony fish appear in the sea

SILURIAN
Ammonites in the sea; plants create soils; first insects and spiders appear on the land. Closes with mass extinction event of unknown cause

ORDOVICIAN-SILURIAN EXTINCTION 445 MYA

PALEOZOIC

ORDOVICIAN
Jawed vertebrates appear, trilobites at their richest; plants and fungi colonise the land. Glaciation occurs along with climate change

PHANEROZOIC EON

PALEOZOIC

CAMBRIAN
Animal life rapidly diversifies. All major animal groups appear, plus some indefinable forms

CAMBRIAN EXPLOSION

PROTEROZOIC EON

PROTEROZOIC EON 2,500–542 million years ago
The atmosphere becomes oxygenic and the first multi-celled animals emerge

ARCHAEAN EON 3.8–2.5 billion years ago
prokaryotic life begins; first stromatolites

HADEAN EON 4.5–3.8 billion years ago
Formation of Earth and Moon

Single-celled/prokaryotic life

550 MYA Oxygen levels reach present day
575 MYA Ediacaran period: Enigmatic soft-bodied multicellular organisms appear
600 MYA Snowball Earth

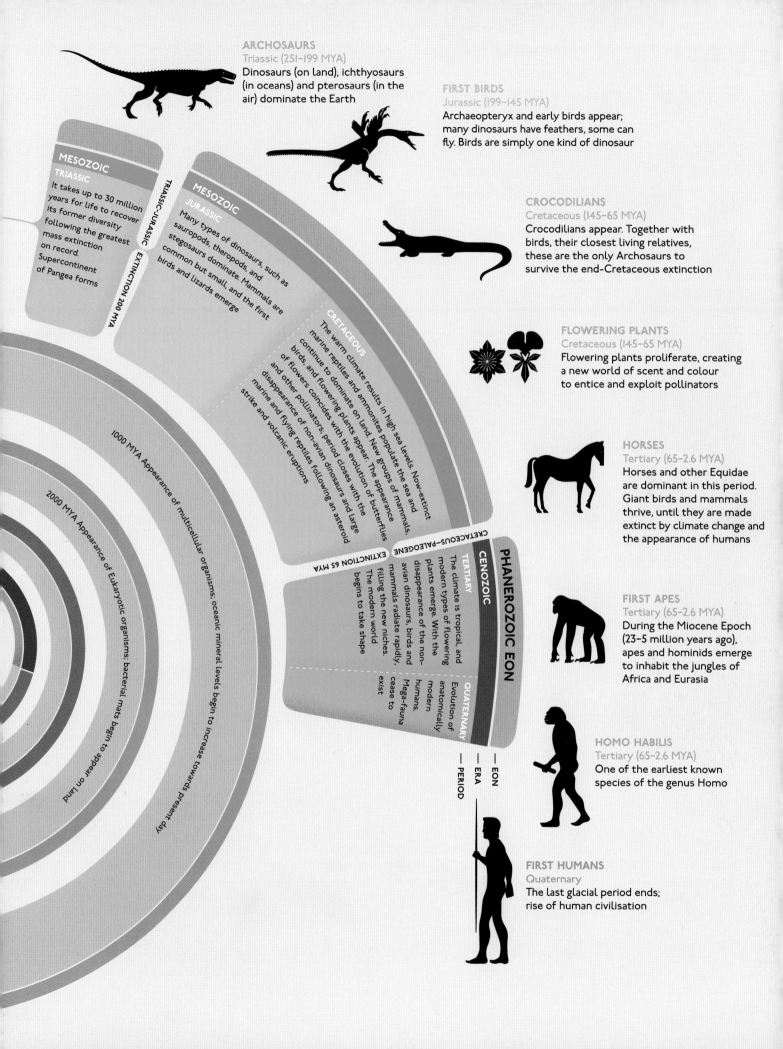

## ARCHOSAURS
### Triassic (251–199 MYA)
Dinosaurs (on land), ichthyosaurs (in oceans) and pterosaurs (in the air) dominate the Earth

## FIRST BIRDS
### Jurassic (199–145 MYA)
Archaeopteryx and early birds appear; many dinosaurs have feathers, some can fly. Birds are simply one kind of dinosaur

## CROCODILIANS
### Cretaceous (145–65 MYA)
Crocodilians appear. Together with birds, their closest living relatives, these are the only Archosaurs to survive the end-Cretaceous extinction

## FLOWERING PLANTS
### Cretaceous (145–65 MYA)
Flowering plants proliferate, creating a new world of scent and colour to entice and exploit pollinators

## HORSES
### Tertiary (65–2.6 MYA)
Horses and other Equidae are dominant in this period. Giant birds and mammals thrive, until they are made extinct by climate change and the appearance of humans

## FIRST APES
### Tertiary (65–2.6 MYA)
During the Miocene Epoch (23–5 million years ago), apes and hominids emerge to inhabit the jungles of Africa and Eurasia

## HOMO HABILIS
### Tertiary (65–2.6 MYA)
One of the earliest known species of the genus Homo

## FIRST HUMANS
### Quaternary
The last glacial period ends; rise of human civilisation

### MESOZOIC
#### TRIASSIC
It takes up to 30 million years for life to recover its former diversity following the greatest mass extinction on record. Supercontinent of Pangea forms

#### TRIASSIC–JURASSIC EXTINCTION 200 MYA

### MESOZOIC
#### JURASSIC
Many types of dinosaurs, such as sauropods, theropods, and stegosaurs dominate. Mammals are common but small, and the first birds and lizards emerge

#### CRETACEOUS
The warm climate results in high sea levels. Now-extinct marine reptiles and ammonites populate the sea and continue to dominate on land. New groups of mammals, birds, and flowering plants appear. The appearance of flowers coincides with the evolution of butterflies and other pollinators; period closes with the disappearance of non-avian dinosaurs and large marine and flying reptiles following an asteroid strike and volcanic eruptions

#### CRETACEOUS–PALEOGENE EXTINCTION 65 MYA

### CENOZOIC
#### TERTIARY
The climate is tropical, and modern types of flowering plants emerge. With the disappearance of the non-avian dinosaurs, birds and mammals radiate rapidly, filling the new niches. The modern world begins to take shape

#### QUATERNARY
Evolution of anatomically modern humans. Mega-fauna cease to exist

### PHANEROZOIC EON

— EON
— ERA
— PERIOD

1000 MYA Appearance of multicellular organisms; oceanic mineral levels begin to increase towards present day

2000 MYA Appearance of Eukaryotic organisms; bacterial mats begin to appear on land

# RETURN OF THE KING

With its vivid orange colour and beautiful markings, the monarch butterfly (*Danaus plexippus*) is a striking example of the simple aesthetic beauty of life. But as is so often the case in the natural world, the superficial beauty of these butterflies is immeasurably enhanced by a deeper scientific understanding of their life cycle and biochemistry, and the reasons for their form and function.

Each year, as autumn approaches across Canada and the northern United States, millions of monarch butterflies begin preparations for an arduous expedition. To survive the harsh northern winter, they embark on one of nature's great migrations, travelling up to 4,000 km to warmer domains in the south. It is a vast distance for such a small and seemingly fragile creature to travel, and requires the birth of a special generation of butterflies. An average adult monarch has a life span of little more than four weeks, but, when faced with the journey south, a 'methuselah generation' emerges; a generation that lives nearly ten times longer than its parents and grandparents.

Living for up to eight months, these butterflies carry with them the privilege of a longer life and the responsibility of carrying their genes through to the following year. As autumn begins in the forests, fields and meadows of the north, preparation starts for travel. The fading of the northern Sun, a result of Earth's journey around the

Sun coupled with the 23-degree tilt of its axis, causes temperatures to fall and food to become scarce. By early September, the young butterflies sense the shortening days and begin to gorge themselves on nectar, laying down extra layers of fat to increase their resilience. When the temperature approaches the very limits of their tolerance, they take flight. This is no random journey south. Covering up to 100 km a day, half a billion monarchs head towards a very specific location. None of them has travelled the route before, yet their destination has remained the same for thousands of years.

Of the many possible solutions to this annual challenge, the monarch butterflies have evolved into skilled navigators, using time and a star as their guide. From their starting point east of the Rocky Mountains, they journey across the great plains of the central United States into the damp humidity of the south. Along the way they face the same dangers as all long-distance travellers; illness and infection, bad weather and storms are a constant danger, and predatory birds will pick off thousands before they come close to completing their annual voyage.

But every year, despite the daunting distance and difficulties, millions of monarchs arrive in a single small area of evergreen forest in the heart of central Mexico. Populations of

**LEFT:** In early September each year, monarch butterflies gather in their millions east of the Rocky Mountains before migrating south to the evergreen forests of central Mexico.

**BELOW:** This magnified image of the head of a butterfly clearly shows its long, segmented antennae, its two segmented eyes, and its tightly coiled proboscis – the three most important sensory organs.

**USING SPECTRAL GRADIENTS TO FIND THE POSITION OF THE SUN:** Long wavelengths (green light) dominate the solar hemisphere, and shorter wavelengths (violet) dominate the anti-solar hemisphere.

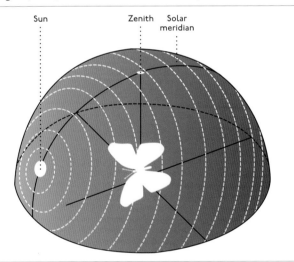

Sun · Zenith · Solar meridian

monarchs that were living west of the Rockies will have made a similar, though shorter, voyage to safety in southern California.

The monarchs navigate like eighteenth-century explorers, using the position of the Sun in the sky and an internal clock to guide them. Taking a southerly bearing using the Sun is simple if you know the time. At noon in the northern hemisphere, the Sun will always be due south. This can be taken as a definition of noon. You can take a southerly bearing at other times of day if you have a watch. Point the hour hand at the Sun, and the line halfway between the hour hand and the 12 o'clock mark will point due south. The monarchs use a sophisticated version of this technique – known as a time-compensated Sun compass – to maintain their southerly orientation during their migration.

The butterflies measure the position of the Sun using their sophisticated eyes, which can detect the polarisation of sunlight, enabling them to 'see' the position of the Sun, even through cloud. They are also thought to use 'spectral gradients', whereby the precise mixture of colours in any given patch of sky depends on how close it is to the Sun. This is due to the way that different wavelengths of sunlight scatter in the atmosphere, an effect that is most familiar in the reddening of the sky at sunset and sunrise.

The nature of the monarch's clock is more elusive. Biological clocks are ubiquitous in nature and thought to be a very ancient evolutionary invention. Circadian rhythms, which require the beating of an internal biological clock, are found in every corner of the biosphere, from the most complex of mammals to the simplest of bacteria. It is possible that biological clocks could have emerged as a form of protection against the destructive effects of the Sun's radiation. An organism's DNA is most exposed to damage at the point of replication, so restricting cell division to the hours of darkness would have been advantageous. This requires a clock that is synchronised to the rotation of the Earth.

Until recently it was assumed that, in common with other animals, the monarch's clock must reside in the brain. But an experiment conducted by neurobiologists at the University of Massachusetts Medical School in 2009 revealed that it is instead located in the delicate structure of the antennae. The reason for this unusual location is not known. Timing information from the antenna clock is combined with information on solar position from the eyes in dedicated regions deep within the butterflies' tiny brains, and this allows them to maintain a southerly bearing on their journey to central Mexico.

For the next five months, a handful of Mexican valleys are home to a billion butterflies, clustering on the firs in such numbers that the forests are painted with a magnificent orange glow. The monarch migration is a powerful example of the way that an organism's home is not a fixed place, but rather a set of conditions that enable it to survive. If those conditions change, it may be necessary to move.

The monarch is an evocative example of a deep truth in biology. The form and function of an animal cannot be understood in isolation. The monarch's behaviour and biochemistry are intimately connected with its habitat, the behaviour of countless other animals and plants, and the constantly shifting seasons driven by the dynamics of the Solar System. I find it simultaneously trivial and wonderful to observe that there would be no monarch butterflies as we know them if our planet's spin axis were not tilted; there would be no seasons, and no evolutionary imperative for migrations. The reason for the tilt is undoubtedly pure chance – a relic of our planet's formation and history stretching back over 4.5 billion years. Jupiter and Mercury have virtually no tilt, while Uranus rotates on its side.

This poses a series of interesting questions: What are the factors that make Earth a home to such a bewilderingly rich and complex ecosystem? What is the minimal set of ingredients necessary for life to evolve, and how widespread are these ingredients in the Universe beyond Earth? Is the emergence of complex living things such as monarch butterflies, fir trees and human beings an inevitable consequence of the laws of physics, or does it rely on a home whose existence is so improbable that Earth and its living ecosystem is a rare, even unique, corner of the Milky Way galaxy, itself one of billions of galaxies in the observable Universe? ◉

**RIGHT:** The behaviour and biochemistry of monarch butterflies cannot be understood in isolation either from their habitat or from the shifting seasons.

# A VERY
# SPECIAL HOME

It is difficult to do justice in a few short paragraphs to Mexico – or, as it is more correctly known, the United Mexican States. An intense and colourful country of contradictions, it is both welcoming yet occasionally frightening, peaceful yet troubled. It has a striking veneer of colonial architecture and customs, but the magnificent architecture of its great indigenous civilisations is intact and imposing, and their ancient mythology makes a vibrant contribution to twenty-first-century global culture. What schoolchild isn't fascinated by the Aztecs, and which New Age conspiracy theorist doesn't read infinitely too much into the Mayans' fascination with the creation of complex and far-reaching calendars?

Physically, Mexico covers almost 2 million sq km and is home to 112 million people. Bordering the United States of America to the north, the Pacific Ocean to the south and west, the Gulf of Mexico to the east, and Guatemala, Belize and the Caribbean Sea to the southeast, it is a land of tremendous geological and climatic variation – from lowland rainforests to pine savannahs; from fertile grasslands to high volcanic mountain ranges. Its position – straddling the Tropic of Cancer and bounded by two of the world's great oceans – also makes it one of the most biodiverse countries on Earth. Even though it

**HOT SPOTS OF HIGH ENDEMISM AND SIGNIFICANT THREAT OF IMMINENT EXTINCTION**

*Mexico is one of the most biodiverse countries on Earth. Even though it covers only 1 per cent of the land area of our planet, it is home to over 200,000 different species – 10 per cent of Earth's bank of life.*

**BELOW:** Cenotes (a type of sinkhole) mark the edge of a massive crater, formed 65 million years ago when an asteroid, measuring some 10 km in diameter, smashed into Earth.

**RIGHT:** The Mexican beaded lizard (*Heloderma horridum*) is just one of the 707 species of reptile known to exist in Mexico.

**ABOVE:** The cenotes of the Yucatan peninsula contain remarkably clear water, which has been filtered through the porous limestone above over many thousands of years.

**ABOVE:** The ocellated turkey (*Meleagris ocellata*) resides primarily in the rainforests of Mexico's Yucatan peninsula. Only the male – shown here – has such striking plumage.

covers only 1 per cent of the land area of our planet, it is home to over 200,000 different species – at the last count, 10 per cent of Earth's bank of life. There are 707 species of reptiles, 438 species of mammals, 290 species of amphibians and over 26,000 species of flora. This is why we chose Mexico to tell the story of the ingredients that make our world such a comfortable home for life.

We began filming in the tropical rainforests of the Yucatan peninsula, where accessible water resources can be unexpectedly scarce. Large areas of the Yucatan are devoid of rivers and streams because the bedrock, composed mainly of limestone, is porous. There is a large subterranean source of fresh water, however, contained in a complex, stratified aquifer. Fortunately for the occupants of the peninsula, this underground water source is easily accessible through a series of sinkholes known as cenotes. The cenotes lead into vast networks of subterranean caverns dissolved out of the limestone over many thousands of years and flooded by the clean waters of the aquifer. The Mayans built their civilisation around cenotes, many of which lie in a strange, semicircular arc centred on a small village called Chicxulub. They mark out the edge of a giant crater, formed 65 million years ago when an asteroid 10 km in diameter smashed into Earth. Known as the Cretaceous-Paleogene extinction event (or the K-T extinction), this impact is the most widely accepted theory for the cause of the mass extinction of the dinosaurs.

The water in the cenotes is exceptionally clear because it is filtered slowly through the porous rocks of the Yucatan before emerging after thousands of years to flood this subterranean world. Diving into the clear darkness of these underground wells is a unique experience and a welcome respite from the heat and insects of the forest. As you journey deeper into the cave systems, the sunlight fades to darkness but an abundance of life can still be found. This is typical of what we find in even the most extreme conditions on the planet. Remove light, heat, soil, plants, insects, and even oxygen, and life still thrives. But one ingredient is, as far as we know, absolutely essential for life to exist. ◉

*The Mayans built their civilisation around cenotes, many of which lie in a strange, semicircular arc centred on a small village called Chicxulub.*

**ABOVE:** Cenotes contain an abundance of life, and taking a dive into their crystal-clear, yet dark, world is a unique experience.

# SIMPLE BUT COMPLEX

**W**ater arguably exhibits the most complex behaviour of any known substance. This may come as a shock, because the ubiquitous familiarity of its chemical signature – $H_2O$ – is the stuff of the most basic of classroom chemistry lessons. Yet this familiarity hides a deep complexity that we are only now beginning to understand. The complexity doesn't lie in the structure of water molecules themselves of course: each molecule is made of three atoms – two hydrogen atoms and one oxygen atom. From chemistry lessons gone by, you might recall that the two hydrogen atoms are covalently bonded to a single atom of oxygen. Oxygen has eight electrons around its nucleus, six of which are in the outer shell; these are known as 'valence electrons'. Four of these are paired together, leaving two lone electrons that would dearly like to pair up with electrons from other atoms.[1] Each hydrogen atom has a single electron, which it readily shares with the electron-hungry oxygen, and the result is a molecule of water.

However, this simple tetrahedral arrangement of a central oxygen atom surrounded by two pairs of electrons and two hydrogen atoms is deceptive, because the structure allows for tremendously complex behaviour when water molecules come together in large numbers. And, as we shall see, this unique behaviour may well make water a prerequisite for the existence of life, not only on Earth, but anywhere in the Universe. Perhaps unsurprisingly, given its dominance in our lives, scientists have been attempting to unlock its secrets for over three hundred years. ◉

[1] *For those who don't like such anthropomorphic language, it is energetically favourable for electrons with opposite spins to pair up in the available energy levels around a nucleus, and there are four available upper energy levels around the oxygen nucleus for the six electrons to occupy.*

---

**MOLECULAR GEOMETRY:** Tetrahedral electron pair geometry

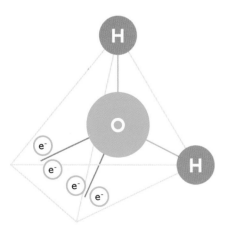

---

**RIGHT:** We often take water for granted, yet it is a remarkably complex substance, and without it there would be no life, not only on Earth, but anywhere in the Universe.

# THE HISTORY OF THE EXPLORATION OF WATER

In the eighteenth century, Europe was full of inquisitive men attempting to unlock the secrets of the natural world, and Henry Cavendish was certainly one of the most eccentric. It is said that he was unable to bring himself to interact with women outside of his family at all, communicating with his female servants by written notes and sneaking around his own house using a specially constructed staircase so as to avoid his housekeeper. His isolation was so extreme that he often kept his experimental findings secret, not publishing or sharing his research with anyone. Such was the extent of this secretiveness that it was only many years after his death that the true breadth of his discoveries became apparent.

Cavendish was a follower of phlogiston theory – a widely held belief that had its roots in alchemy. The theory suggested the existence of an element thought to be contained within all combustible material, called 'phlogiston'. By the middle of the eighteenth century the theory had been widely discredited, yet Cavendish continued to see worth in it, and attempted to incorporate it into many of his observations. To modern ears, this makes his terminology sound rather eccentric, but his contribution to our understanding of the natural world was extraordinary, not least in his early work on the chemical properties of water.

In a series of experiments, Cavendish produced and isolated a gas by reacting hydrochloric acid with metals such as zinc, iron and tin. In doing so, he became the first person to identify hydrogen in the laboratory. He referred to this new gas as 'flammable air' in his poetically named paper 'Factitious Airs', published in 1766. Cavendish went on to show that hydrogen reacted with another gas, which he termed 'dephlogisticated air', to produce water. This gas was oxygen. His experiments with flammable air eventually led him to the first determination of the composition of Earth's atmosphere – one part dephlogisticated air (oxygen) and four parts 'phlogisticated air' (nitrogen). There is something quite instructive in Cavendish's approach to science. Even though his devotion to the phlogiston theory was wayward, to say the least, he did not allow his theoretical prejudice to contaminate his experimental results. This is why he was able to make genuine discoveries while holding at least some views about his subject that were flat-out wrong. That is the mark of a great experimental scientist!

## ELECTROLYSIS OF WATER

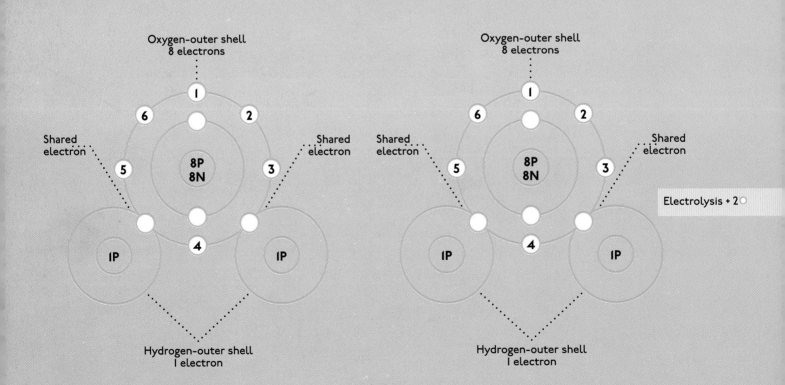

Oxygen-outer shell
8 electrons

Shared electron

Shared electron

8P
8N

IP

IP

Hydrogen-outer shell
I electron

Oxygen-outer shell
8 electrons

Shared electron

Shared electron

8P
8N

IP

IP

Hydrogen-outer shell
I electron

Electrolysis + 2

**P** Proton
**N** Neutron

We owe the modern names for the elemental building blocks of water – hydrogen and oxygen – to Antoine Lavoisier, one of the greatest of the pioneering eighteenth-century chemists. Great though he undoubtedly was, however, he made a fundamental error in naming these two elements that persists to this day.

He named hydrogen, entirely appropriately, from the Greek 'hydro' (meaning water) and 'genes' (meaning creator). Oxygen, however, with its Greek root of 'oxys' (meaning acid), incorrectly suggests that oxygen is a component of all acids. It would have been more accurate to call hydrogen 'oxygen', in that the majority of common acid-base chemical reactions involve the transfer of protons, which are the nuclei of hydrogen. But Lavoisier's names have stayed with us, so oxygen will forever be 'the acid giver', which it isn't.

By 1804, the final elemental description of water was given in a paper by the French chemist Joseph Louis Gay-Lussac and the German naturalist Alexander von Humboldt. Together, they demonstrated that water consisted of two volumes of hydrogen to one of oxygen, and thus gave the world the most widely known of all chemical formulae: $H_2O$. If Lavoisier had got it right, we'd call water $O_2H$ rather than $H_2O$. Such is history.

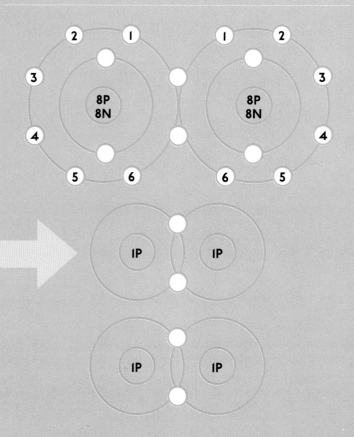

## MR BELL'S GUIDE TO THE ELECTROLYSIS OF WATER

Everybody has a teacher whose very essence, usually distilled from endearing eccentricity, remains forever imprinted on their consciousness. I had such a teacher, and his name was Sam Bell. He used to gaze out of the thin windows over the playing fields on a darkening Oldham afternoon and growl strangely in a thick Yorkshire accent about how he could still see the old games master, Pinky Green, ringing a bell, before launching a board duster across the varnished benches towards a boy's head. He'd undoubtedly face disciplinary proceedings for that today, but it made chemistry enjoyable to an 11 year old. Mr Bell's trademark technique was to drill chemical reactions into your brain with an indelible power usually reserved for poetry. 'The ploughman homeward plods his weary way, and HYYDR'GEN burns with a squeaky POP!' This works, as I discovered when, half a lifetime later, I settled down with a car battery in front of a waterfall in central Mexico to explain the electrolysis of water to a television camera.

'Electrons enter the water at the cathode, where a reduction reaction takes place, releasing hydrogen gas which burns with a squeaky pop in air. Oxidation occurs at the anode, producing oxygen, which rekindles a glowing splint.' Perfect.

Cathode (reduction):  $2 H_2O + 2e^- \longrightarrow H_2 + 2 OH^-$
Anode (oxidation):  $4 OH^- \longrightarrow O_2 + 2 H_2O + 4e^-$

The point of all this, which will be important when we come to discuss photosynthesis later on, is to demonstrate that it takes a large amount of energy to split water into hydrogen and oxygen. This is because oxygen really wants to acquire the two extra electrons necessary to fill its outer shell, and hydrogen is a relatively easy place to get them. This in turn means that water is a very stable molecule, and it therefore takes a lot of effort – in this case the power of a car battery – to split it apart. But, really, the point is to show that I thought my chemistry teacher was brilliant. ◉

### ELECTROLYSIS OF WATER

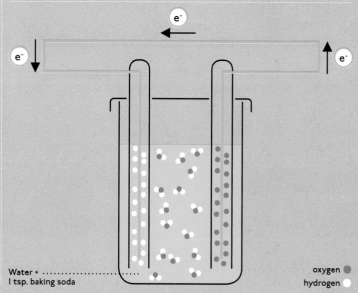

Water + ........................
1 tsp. baking soda

oxygen ◉
hydrogen ○

# WATER, WATER EVERYWHERE...

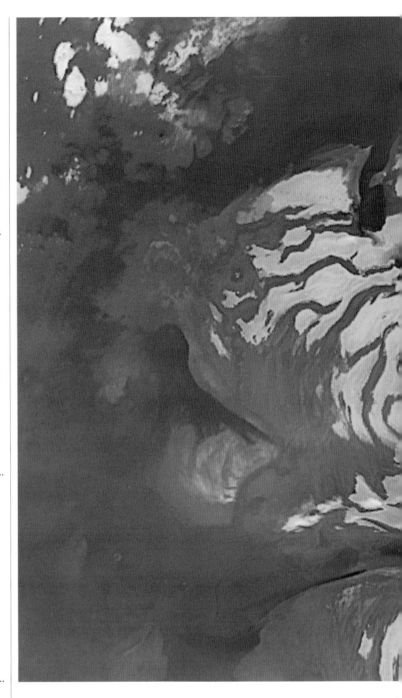

**ABOVE:** Water on Mars? This composite image, taken by NASA's Mars Reconnaissance Orbiter, shows the polar ice cap of the 'red planet'. The ice cap is believed to be made of ice and dust deposits.

Earth, the small blue planet, is unique within the Solar System in having liquid water on its surface today. Indeed, from space Earth is a water world, with 71 per cent of its surface covered by the liquid. This uniqueness is the result of Earth's size and position in the Solar System, and not the scarcity of the life-giving molecule itself. Despite the fact that we have yet to discover another water world like our own, we know that, across the vast expanses of space, the seas, lakes and rivers of planet Earth are just a drop in the cosmic ocean – our Universe is literally awash with water molecules, and everywhere we look in space we can see that the Universe is wet.

This shouldn't be surprising when you consider that hydrogen and oxygen are two of the most abundant atoms in the Universe. Hydrogen forms 74 per cent of all the elemental mass. The second-lightest element, helium, comprises 24 per cent. These two elements dominate because they were formed in the first few minutes after the Big Bang. Oxygen is the third most abundant element in the cosmos, at around 1 per cent by mass. Most of the rest is carbon; all the other elements are present in much smaller quantities. All of the oxygen and carbon atoms in the Universe today, including all of those

*All of the oxygen and carbon atoms in the Universe today ... were produced in the cores of stars by nuclear fusion and scattered out into space as the stars died.*

in your body, were produced in the cores of stars by nuclear fusion and scattered out into space as the stars died. Apart from helium, which is satisfied with its full inner shell of two electrons, these atoms have an affinity for each other because of their desire to pair up their solitary electrons. As a result, they tend to form molecules. After the hydrogen molecule ($H_2$) and carbon monoxide (CO), water is the third most common molecule in the Universe.

Much of this interstellar ocean is created during the formation of stars. There are over 400 billion stars in our Milky Way galaxy alone, and each time a new star is born a chain of events leads to the production of water. Stars are formed when an interstellar cloud of gas collapses under the force of gravity. As the gasses fall inwards, they heat up until nuclear fusion is initiated. This process of collapse, followed by ignition, creates a powerful outward burst of gas and dust.

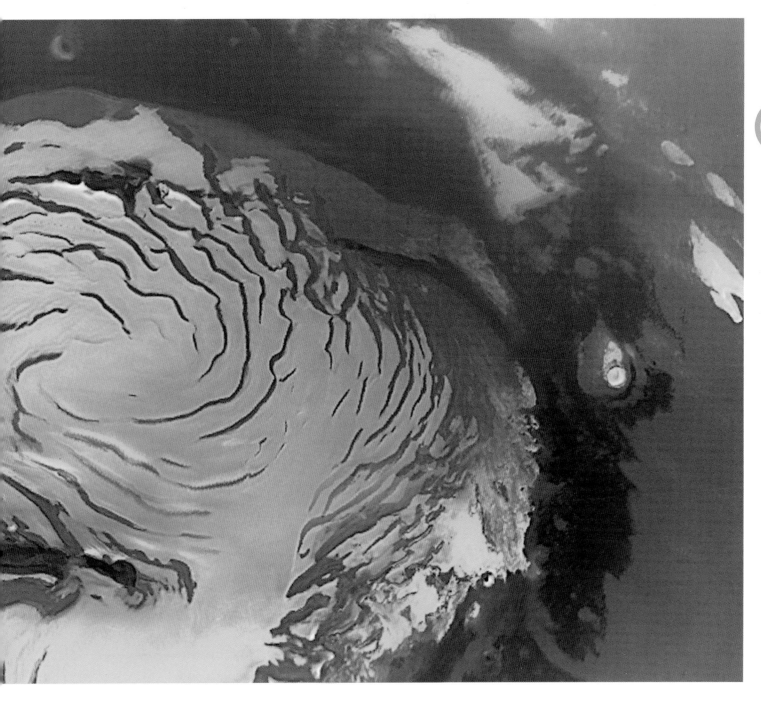

When this material hits the surrounding molecular cloud, already rich in oxygen from previous stellar deaths, the plentiful hydrogen and oxygen can combine, producing water.

On 22 July 2011, a team of astronomers from NASA's Jet Propulsion Laboratory and the California Institute of Technology (Caltech) announced the discovery of the largest, most distant reservoir of water ever detected. A gigantic cloud of $H_2O$, containing 140 trillion times more water than all of Earth's oceans combined, was sighted over 12 billion light years away from Earth. It surrounds one of the most evocative and powerful objects in the Universe: a quasar with the catchy name APM 08279+5255. This active galaxy harbours a black hole 20 million times more massive than the Sun. The star systems and gas spiralling into this voracious monster release a power output equivalent to 1,000 trillion suns as they slide down the sheer space-time slopes. This generates a shock wave on a galactic scale, forcing hydrogen and oxygen molecules together in unimaginable numbers to produce a giant reservoir of water. The scale of the find is extraordinary, but so is its age. Since the light from the quasar took over 12 billion years to reach Earth, we are seeing the Universe as it was less than 2 billion years after the Big Bang. This reservoir is therefore very ancient indeed, and the discovery proves that life-giving water is not only abundant, but has been present in the Universe for a large fraction of its lifetime.

Water was there from close to the beginning of time, and the Universe is full of it. Our galaxy, the Milky Way, is also full of it, although it's relatively dry compared to APM 08279+5255, with only around 350 billion times more water than Earth. This interstellar reservoir was part of the cloud that, 4.5 billion years ago, condensed into our Solar System and formed the oceans and rivers that cover our blue planet today. ◉

# WATER:
# THE ESSENTIAL INGREDIENT

Water runs deep in the living world. It is the principal ingredient of every living thing on the planet. On Earth, where there is water, there is life. In fact, water is so central to life on Earth that many astrobiologists believe it will be central to any life, anywhere in the Universe.

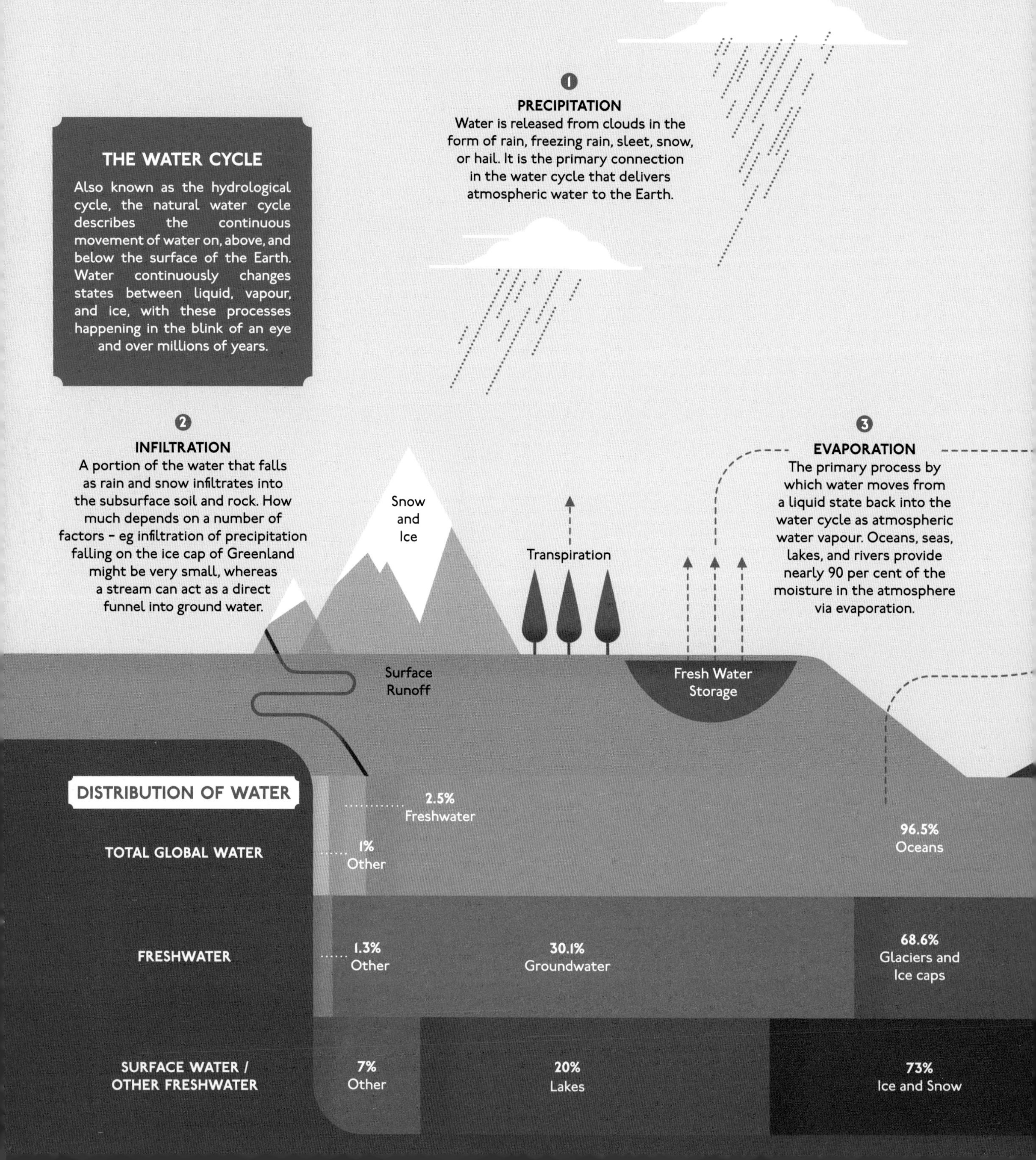

## THE WATER CYCLE

Also known as the hydrological cycle, the natural water cycle describes the continuous movement of water on, above, and below the surface of the Earth. Water continuously changes states between liquid, vapour, and ice, with these processes happening in the blink of an eye and over millions of years.

### ❶ PRECIPITATION

Water is released from clouds in the form of rain, freezing rain, sleet, snow, or hail. It is the primary connection in the water cycle that delivers atmospheric water to the Earth.

### ❷ INFILTRATION

A portion of the water that falls as rain and snow infiltrates into the subsurface soil and rock. How much depends on a number of factors – eg infiltration of precipitation falling on the ice cap of Greenland might be very small, whereas a stream can act as a direct funnel into ground water.

### ❸ EVAPORATION

The primary process by which water moves from a liquid state back into the water cycle as atmospheric water vapour. Oceans, seas, lakes, and rivers provide nearly 90 per cent of the moisture in the atmosphere via evaporation.

Snow and Ice

Transpiration

Surface Runoff

Fresh Water Storage

## DISTRIBUTION OF WATER

| | | | |
|---|---|---|---|
| **TOTAL GLOBAL WATER** | 1% Other | 2.5% Freshwater | 96.5% Oceans |
| **FRESHWATER** | 1.3% Other | 30.1% Groundwater | 68.6% Glaciers and Ice caps |
| **SURFACE WATER / OTHER FRESHWATER** | 7% Other | 20% Lakes | 73% Ice and Snow |

## ⑤ WATER IN THE ATMOSPHERE

Clouds are the most visible manifestation of atmospheric water, but even clear air contains water – in particles that are too small to be seen.

## ④ CONDENSATION

The process by which water vapour in the air is changed into liquid water. Condensation is crucial to the water cycle because it is responsible for the formation of clouds. These clouds may produce precipitation, which is the primary route for water to return to the Earth's surface within the water cycle.

## HOW EARTH GOT ITS WATER

### COMETS

It has long been thought that comets, which are full of ice and organic compounds, brought a large proportion of water to the Earth. However, not all comets have the same composition of water as that found on Earth's oceans.

### METEORITES

Meteorites may account for the wet Earth we inhabit today. Asteroids and some meteorites contain the right ratio of heavy to regular water, and primitive meteorites bear the closest resemblance to the primordial components of the terrestrial planets.

## WATER CONTENT

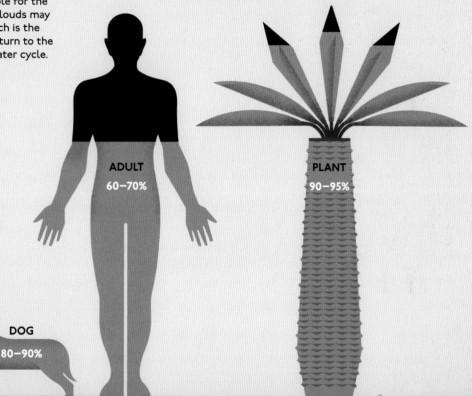

**ADULT**
60–70%

**PLANT**
90–95%

**DOG**
80–90%

**BABY**
78%

**JELLYFISH**
95%

**FISH**
80%

**DEEP SEA CRAB**
86–90%

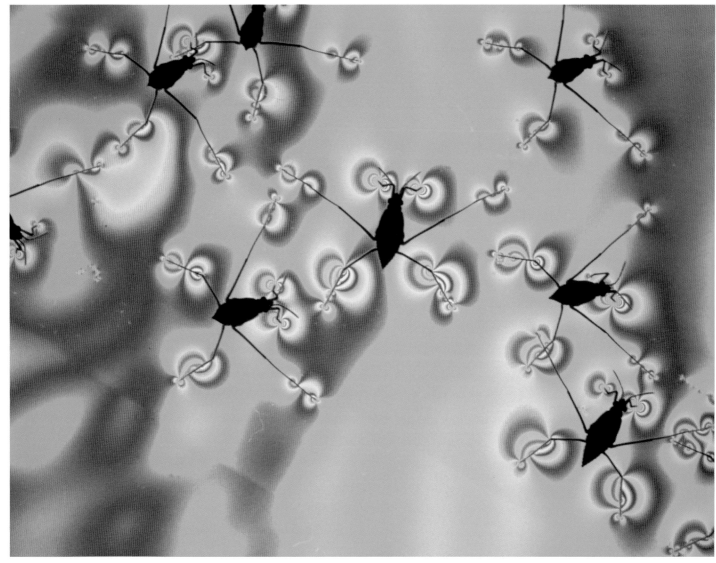

# WALKING ON WATER

O f all the creatures on Earth, few exploit the unique characteristics of water as overtly as the family of insects known as Gerridae. You may also know them as pond skaters, water striders or Jesus bugs.

Gerridae are successful and vicious killers, piercing the body of a captured spider or fly with a specially adapted mouthpart and finishing it off by sucking out its insides. The 1,700 known species around the world are found in a large range of water habitats, from the pond in your back garden to slow-flowing rivers in the deepest recesses of the Mexican jungle. But it is not their killing methodology or diversity that makes these animals so interesting to school children and a physicist dabbling in biology; it is their ability to walk on water, like Jesus. Next time you look at a common pond skater, you'll be observing a creature that exhibits an exquisitely balanced relationship between its anatomical features and the physical properties of water, because the Gerridae is beautifully adapted to life at the interface between water

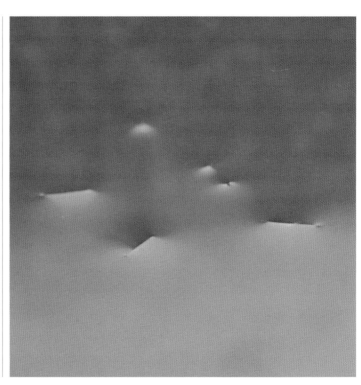

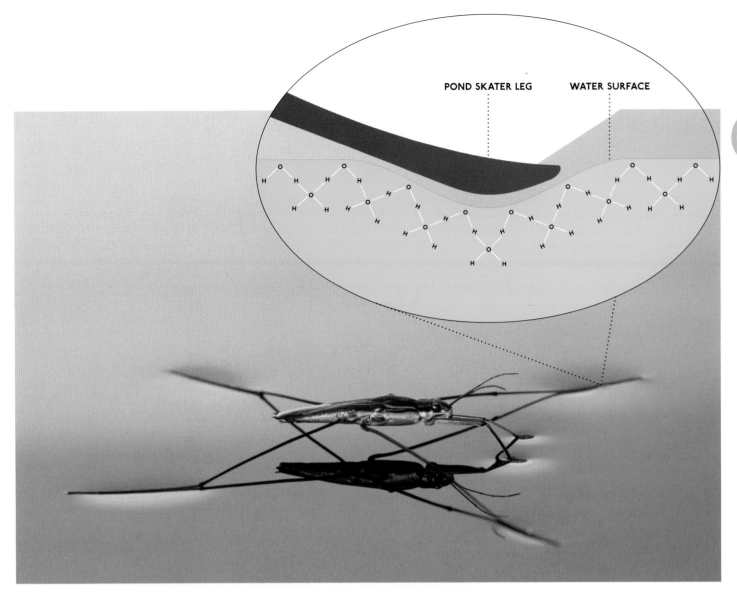

POND SKATER LEG    WATER SURFACE

*Look at a common pond skater, and you'll be observing ... an exquisitely balanced relationship between its anatomical features and the physical properties of water.*

**TOP LEFT:** Pond skaters can walk on water. However, it is not just their anatomical adaptations, but also the physical properties of water, that enable them to occupy this unique environmental niche.

**ABOVE:** The pond skater's back pair of legs spreads the animal's weight over a wider area, while the middle pair propels it through the water.

**LEFT:** This image of pond skaters was taken from underneath the water, looking upwards. The strong bonds between the water molecules help to prevent the insects from breaking the surface.

and air. Its short front legs are used for capturing prey, while its middle legs propel it through the water. Its back legs are long and slender, spreading the animal's weight over a larger surface area. These gangly appendages contribute to the pond skater's ability to walk on water, but alone they would not be enough to keep it afloat. Every square millimetre of its body is covered with a cohort of tiny hairs that increase the surface area still further. These hairs are also hydrophobic, making the whole animal water-resistant. Without this adaptation, a single drop of rain would be enough to weigh the creature down and sink it below the surface. Even if the animal is pushed under, the tiny water-repelling hairs trap air, adding buoyancy and returning the creature to the surface. All of these anatomical features combine to allow the pond skater to live out its life in this unique environmental niche, moving around the water's surface at speeds of up to 1 m per second – remarkably fast for such a small creature. Yet all these clever adaptations alone would not keep a pond skater afloat if it wasn't for the especially strong bonds that exist between the water molecules themselves, and it is ultimately these bonds that make water so vital for life. This is why we chose this common but fascinating little animal as our introduction to the wonder of water. ◉

# TREETOPS TO TEARDROPS: THE MAGIC OF HYDROGEN BONDS

**RIGHT:** Water bends towards a Perspex rod that has been rubbed against a fleece. The rod has an electric charge, caused by the rubbing, and the water bends because of its own polar nature.

Water is colourless, tasteless, odourless and has a simple chemical formula – $H_2O$ – but this simplicity is deceptive, because the geometry of the water molecules themselves means that their collective behaviour is tremendously subtle and complex. The diagram below shows a series of different molecules, each consisting of hydrogen atoms covalently bonded to different elements. The simplest is hydrogen fluoride, which forms a linear structure as there are only two atoms – one of fluorine and one of hydrogen. Fluorine bonds with only one hydrogen atom because it has only a single electron in its outer shell. Water has two electrons available for bonding, but it also has two pairs of electrons sitting inertly in its outer shell. Inert they may be, but they still have to 'fit' somewhere, and their presence means that water molecules are not linear. The hydrogen atoms sit on one side of the oxygen, at an angle of 104.45°.

This has a very important consequence. Electrons are negatively charged, and the nuclei of hydrogen, being single protons, are positively charged. Water's angled geometry means that the region surrounding the hydrogen atoms has a slight positive charge, and the region away from the hydrogen atoms has a slight negative charge. This means that water is a 'polar' molecule – one side is slightly negatively charged, and the other is slightly positive, although the molecule itself remains electrically neutral. This is the reason for the unexpected behaviour of water in a classic school science experiment. Take a Perspex rod (one of those perplexing objects found in every science laboratory but nowhere else) and rub it against a fleece. This gives the rod an electric charge, in much the same way that you might get charged up by walking across a carpet and discharged uncomfortably by grabbing a door handle. If you move the rod next to water flowing out of a tap, the stream of water will bend because the positive and negative sides of the water molecules are either attracted to or repelled by the electric charge on the rod.

It is the polar nature of water molecules that gives this seemingly innocuous liquid an array of complex properties so vital for life on Earth. The molecules are not only attracted to or repelled by external electrically charged objects, they also attract each other, forming weak bonds known as hydrogen bonds. Water isn't the only liquid to do this – hydrogen fluoride and ammonia also exhibit hydrogen

**HYDROGEN BONDS**

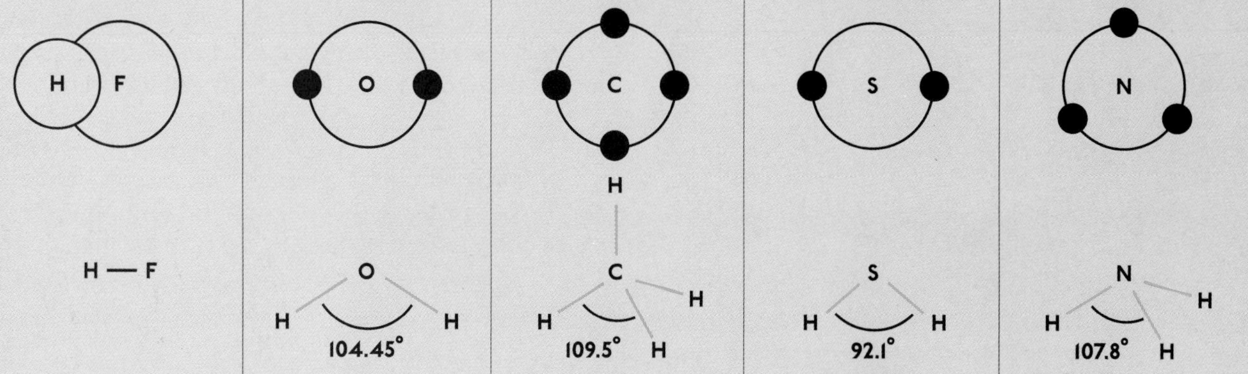

| | HYDROGEN FLUORIDE | WATER | METHANE | HYDROGEN SULPHIDE | AMMONIA |
|---|---|---|---|---|---|
| MASS | 20.01 | 18.01 | 16.04 | 34.08 | 17.03 |
| MELTS | -83.6°C | 0°C | -182°C | -82°C | -77°C |
| BOILS | 19.5°C | 100°C | -162°C | -60°C | -33.34°C |
| DIPOLARITY | 1.86D | 1.85D | 0D | 0.97D | 1.42D |
| HYDROGEN BONDING | YES | YES | NO | NO | YES |

bonding for the same reason – they have a negative and positive side to them because of their geometry and the distribution of the electrons around their component atoms.

One of the most immediate consequences of hydrogen bonding is a dramatic rise in the boiling point of these substances. Methane, which is a symmetric molecule because it has four hydrogen atoms surrounding its central carbon atom (see diagram), is not polar and does not exhibit hydrogen bonding. This means that methane molecules are only very weakly bonded together in the liquid state, and it doesn't take much energy to split them apart from each other and turn liquid methane into a gas. This is why the boiling point of methane is a chilly –162°C. Ammonia, on the other hand, with only one hydrogen less than methane and a very similar molecular size and weight, exhibits hydrogen bonding because it is polar, and its boiling point is a fairly warm –33°C, a temperature regularly reached in cold areas on our planet. Hydrogen fluoride is also polar, because of fluorine's voracious appetite for electrons, and it boils at room temperature. And water, of course, boils at 100°C at room temperature and standard atmospheric pressure, because of its strong

hydrogen bonds. We can get an idea of the importance of hydrogen bonds by comparing water to hydrogen sulphide, a very similar molecule in terms of weight and size, but with an atom of sulphur replacing the oxygen atom at its heart. $H_2S$ does not exhibit hydrogen bonding, because the sulphur atom does not drag the electron cloud around it as effectively as oxygen. This is because it has an extra inner shell of electrons shielding the positive electric charge of its nucleus. As a result, $H_2S$ boils at –60°C. Without hydrogen bonding, therefore, there would be no liquid water on the balmy Earth – no oceans, no rivers and lakes, no raindrops and no life.

It is also water's strong hydrogen bonds that explain the pond skater's ability to walk on water. To understand why, it is necessary to think just a little about the nature of chemical bonds themselves. The reason a bond forms, at the most fundamental level, is because it is energetically favourable for it to do so. This means that a clump of water molecules loosely attached to each other by a network of hydrogen bonds is a lower energy configuration than a swarm of water molecules freely whizzing around ignoring each other.

Think about what it means to boil water. You have to put energy into the water to boil it and produce steam; steam is gaseous water, which means that the molecules are whizzing around ignoring each other. When you heat water up, some of the energy goes into breaking the hydrogen bonds between the water molecules. If you have to put energy in to break the bonds, then it must mean that you get energy out by letting the hydrogen bonds re-form and allowing the steam to condense back into water again. This is why steam burns you easily – when it touches your skin and condenses into water, a large amount of energy is released and this hurts! Part of what you are feeling is the energy released as the network of hydrogen bonds re-forms, turning the steam back into liquid.

Because the hydrogen-bonded liquid state of water is a lower energy configuration than the non-hydrogen-bonded gaseous state, this has an interesting effect at the water's surface. Hydrogen-bonding lowers the energy of a collection of water molecules, so every water molecule wants to hydrogen bond to others if it can. The molecules at the surface, however, don't have as many molecules to bond with, as above them there is only air. This means that it is always energetically favourable for water to minimise its surface area; less surface means more hydrogen bonds.

When a pond skater puts its hairy, hydrophobic legs onto the water's surface, it bends the surface and therefore increases the surface area. This increases the energy of the water, which pushes back, trying to flatten its surface and thereby reducing its energy. This force is known as surface tension, and it keeps the pond skater afloat. This is also, by the way, the reason why raindrops are spherical. A sphere is the shape that minimises the surface area of a water drop, and it is therefore the most energetically favourable shape for a collection of water molecules to assume.

Water's high boiling point and surface tension are just the beginning, as far as biology is concerned. Water's polar nature doesn't only allow the formation of hydrogen bonds between water molecules, it also allows it to break up other weakly bonded molecular structures and disperse them. In other words, it is a superb solvent, able to dissolve salts and other nutrients which in turn allows them to be dispersed around the body and made available for chemical reactions to take place. It is also highly structured in its liquid phase. We now know that water behaves more like a gel than a liquid, with complex networks of hydrogen-bonded water molecules forming giant, fleeting structures. These structures, it is thought, play a vital role in the complex biological reactions within cells. In a sense, water acts like scaffolding around which biology can happen. It is known that the activity of proteins depends both on their chemical structure and their precise orientation and shape, and hydrogen bonding between water molecules and the protein molecules plays an important role in orientating these complex molecules so that they can carry out their biological functions correctly.

Water is a fascinating and unique substance – so much so that its influence on biology and its own internal structure, both created by hydrogen bonding, are still extremely active areas of research. This is why it is said that we won't truly understand biology until we understand water. It is also the origin of the strong suspicion, shared by many biologists, that water is one of the essential ingredients for life, not only on Earth, but anywhere in the Universe. ◉

## INTO THE LIGHT

Life on Earth is a dazzling continuum of organisms of dramatically varying sizes and complexities. There are some ingredients that all living things share: water, plus a handful of chemical elements vital for life, such as the constituents of DNA – hydrogen, oxygen, nitrogen, carbon and phosphorus. Other ingredients and conditions are necessary for the particular biosphere we find on Earth today. Without them, human beings would certainly not exist, but whether they are fundamental to the development of complex life is an open question.

Virtually every living thing on the planet today is ultimately powered by sunshine. Every mouthful of every meal has its origins in the Sun, from the fruit and vegetables created by plants that absorb sunlight directly, to the meat and fish that deliver their sunshine second- or third-hand as part of the complex food chain.

It appears today as if the Sun is a truly fundamental ingredient for life, a provider without which life couldn't exist. Yet this intimate relationship with our nearest star is not a simple one. The Sun is a far from benevolent companion. Its radiant rain has a dark side that is as dangerous as it is nourishing, and early in the development of life on Earth it is likely that the Sun was a presence to be avoided rather than cherished. To understand how life transformed its relationship with light, we have to go back billions of years, to a time when life sheltered in the darkness. For many biologists, life on Earth didn't begin in the light, but rather in the darkness of the deep oceans. The transformation of light from threat to food required one of life's most extraordinary inventions: oxygenic photosynthesis. The evolution of this biological process ultimately resulted in the capture of carbon and the release of large amounts of oxygen into the atmosphere, which in turn played a key role in triggering the explosive evolution of life from the simple to the complex and conscious. ◉

**BELOW:** Without light, the process of oxygenic photosynthesis would not be possible. It was this biological process that resulted in the release of oxygen into the Earth's atmosphere.

# A TRAIN JOURNEY THROUGH TIME

As a rule, I don't enjoy filming in jungles. Humidity and DEET combine, in my view, to create discomfort, and it is unfortunate that biodiversity implies lots of animals, some of which are best avoided. It was with some relief, therefore, that the crew and I left the verdant but challenging beauty of the Yucatan behind for the freshening altitudes of the north. With 37 bridges, 86 tunnels and a vertical climb of 2,400 m along 673 km of track, the Chihuahua-Pacific railway is one of the great train journeys of the world. The old train leaves the coastal town of Los Mochis at 6 am, shortly before the reddening sky delivers stripes of warming light into the wooden carriages through slatted blinds. As the train rattles away from the coastal plane, the landscape shifts from towns and villages to pine forests and mountains, and we head inwards and upwards towards Mexico's mountainous interior and the Copper Canyon, a network of gorges to rival its more famous northerly neighbour, the Grand Canyon.

One shouldn't need a reason to ride the Chihuahua-Pacific. It's one of those things in life that's worth doing, simply because it's there. But, being a film crew, we have a reason; to observe the shifting nature of the Sun's light as it arrives at the surface of the Earth. As early morning turns to afternoon and we rise into thinning air, the colours of the world shift from warm reds to harsher blues. These are real, physical changes picked up by our eyes as the quality of the sunlight itself changes owing to its varying path-length through the atmosphere.

Light is an electromagnetic wave; energy from the Sun sloshing back and forth between electric and magnetic fields and driving itself through the vacuum of space at just over 299,792 km per second. Being a wave, light has a wavelength, and it is different wavelengths of light that we see as different colours. Human eyes are sensitive to wavelengths between around 400 nanometres and 700 nanometres (a nanometre is a billionth of a metre). We see 450 nm light as blue, 500 nm light as green, and 600 nm light as orangey-red. Beyond the red, at longer wavelengths, lies the infrared. This is the radiation that we feel as heat, but cannot see. Some animals, such as the pit vipers, have evolved the ability to detect infrared light, but humans need night-vision goggles to do so.

At the short-wavelength end of the spectrum lies the ultraviolet. There are many animals that can see this light,

**BELOW LEFT:** The Chihuahua-Pacific railway has a vertical climb of 2,400 m (8,000 ft) and, as it climbs, the air thins and the colours of the world shift from warm reds to harsher blues.

**BELOW:** The red light of sunrise, seen from the Chihuahua-Pacific railway.

**BELOW MIDDLE:** The train climbs through the Copper Canyon in Mexico's interior.

from birds and bats to insects. Some flowers are intensely colourful in the ultraviolet part of the spectrum, yet their beauty remains hidden to us. The Sun is intensely active in the ultraviolet, but much of it is absorbed by the Earth's atmosphere and never reaches the surface. This is a good thing, because short-wavelength UV is potentially extremely damaging to living things. To see why this is so, it is easier to think of light in another way. Light can also be viewed as a stream of particles called photons. The particulate nature of light was discovered at the turn of the twentieth century by Albert Einstein and others, and was one of the first steps in the development of quantum theory. The 'quanta' of light are known as photons, and the smaller the wavelength of the light, the higher the energy of the photons. High-energy UV photons are like little bullets, smashing into biological molecules with more than enough energy to break them apart. This is the reason why UV light can be dangerous. As the wavelength lengthens, however, the relationship of UV light with life becomes more ambiguous. Long-wavelength ultraviolet light, known as UVB, is beneficial to life (our bodies use it to produce vitamin D), but, just like the shorter wavelengths (UVC), it can also be damaging. UV light certainly poses a challenge to living things.

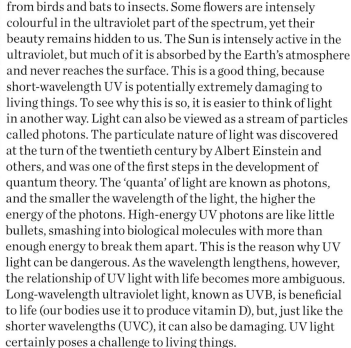

*As early morning turns to afternoon and we rise into thinning air, the colours of the world shift from warm reds to harsher blues.*

As the hours pass on the Chihuahua-Pacific and we rise into the Mexican interior, two things happen: the Sun rises in the sky and the atmosphere becomes thinner. As the Earth's atmosphere absorbs and scatters light of different wavelengths by different amounts, this changes the relative mixtures and intensities of wavelengths of light to which life is exposed. In particular, as the Sun climbs in the sky and the train climbs in altitude, the amount of potentially damaging UVB light rises dramatically. I measured the flux of UVB during the train journey with a small detector called a digital radiometer. In the early morning, with the Sun low in the sky at sea level, I measured a UVB flux of 22 microwatts per square centimetre. This is the power delivered to each square centimetre of a living thing by the ultraviolet light from the Sun. As the train climbed with the rising Sun, the UVB flux climbed to 260 microwatts per square centimetre, because the high-energy UVB photons had less atmosphere to pass through and were therefore less likely to be scattered or absorbed. My body's response to this onslaught is to produce a pigment known as melanin. In simple terms, I get a suntan.

42

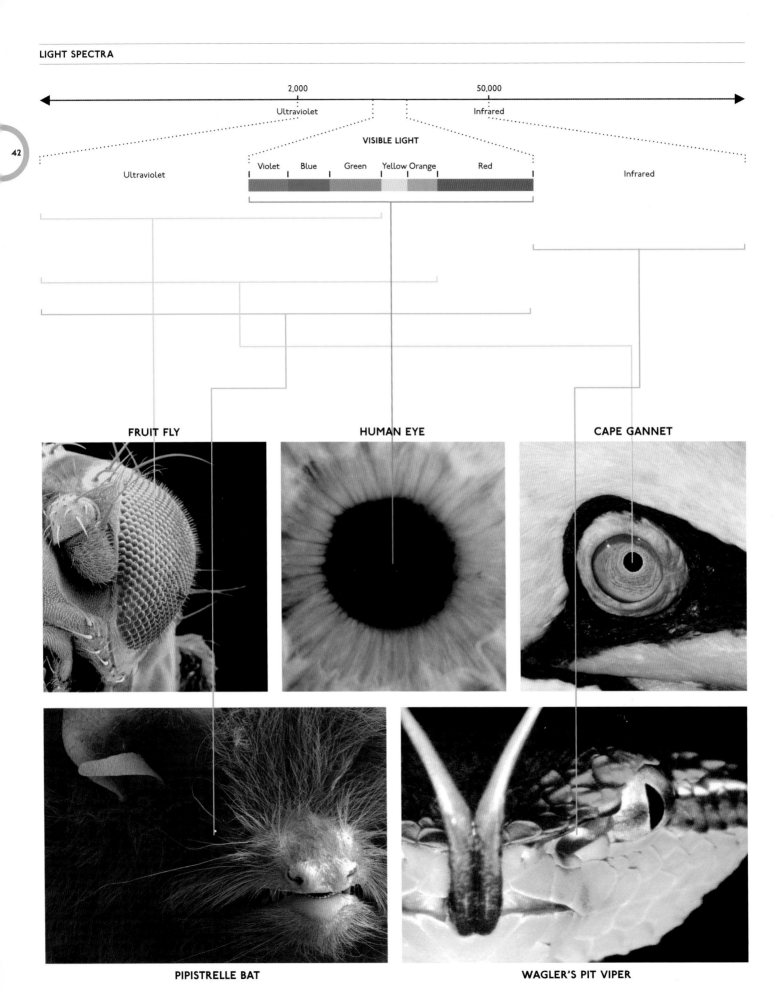

2,000      50,000

Ultraviolet      Infrared

**VISIBLE LIGHT**

Ultraviolet

Violet   Blue   Green   Yellow   Orange   Red

Infrared

**FRUIT FLY**

**HUMAN EYE**

**CAPE GANNET**

**PIPISTRELLE BAT**

**WAGLER'S PIT VIPER**

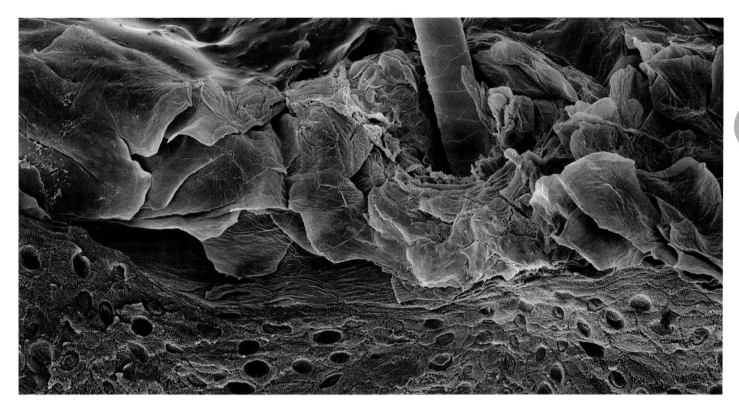

*The use of pigments such as melanin evolved very early in the history of life on Earth – forming a fundamental component of life. They are the way that life interacts with light and protects itself from harm.*

**TOP:** Coloured scanning electron micrograph of a section through human skin. The darker layer contains cells called melanocytes, which produce melanin.

Melanin is found in virtually all animals. In humans, this dark pigment is found in a type of cell under our skin called melanocytes. As high-energy ultraviolet photons rain down on our skin, they have the potential to damage the sensitive molecules that lie beneath. DNA is particularly susceptible to damage from UV, with potentially deadly consequences, and melanin is the first line of defence. The secret to its ability to shield cells from the damaging effects of high-energy photons lies in its molecular structure. Melanin is a complex molecule able to form polymers with varying structures depending on their location in the body. Its active heart, however, is a series of rings of carbon atoms bound together by a sea of mobile electrons. When a high-energy photon from the Sun hits one of the electrons, it doesn't break the molecule apart. Instead, the energy is dissipated in around a pico-second, which is very fast indeed. In a million millionths of a second, the potentially threatening photon has been adsorbed and all its energy has been converted to heat. The melanin molecule survives intact to fight another day. Melanin is so efficient that over 99.9 per cent of the harmful UV radiation is adsorbed in this way, protecting cells from damage.

Melanin in its many forms is ubiquitous in nature; it is even found deep in the human brain, where its function is unknown. Even microorganisms such as bacteria and fungi employ melanin to protect themselves from UV radiation. This suggests that the use of pigments such as melanin evolved very early in the history of life on Earth – forming a fundamental component of life. They are the way that life interacts with light and protects itself from harm. While this would probably have been irrelevant for the very first life forms on Earth, which most likely lived deep in the oceans around hot-water vents, the dangers of UV light would have been one of the first challenges faced by life as first it rose to the ocean surface and then eventually colonised the land. ◉

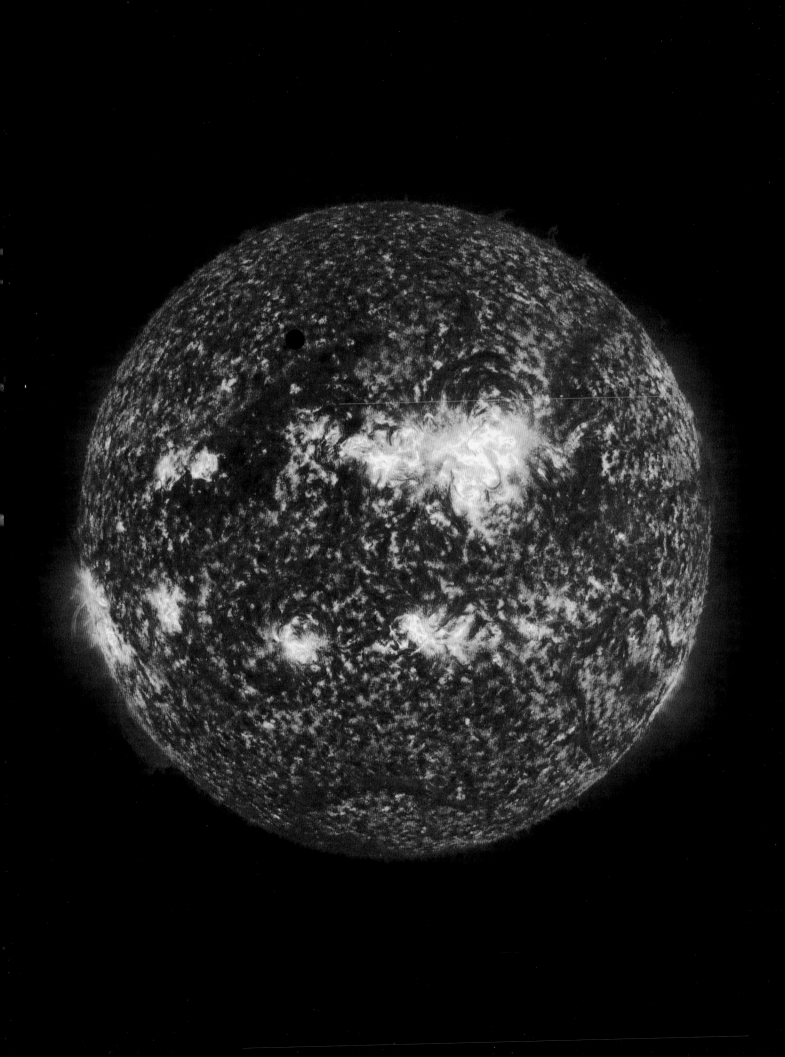

# A CHILD STAR

**LEFT:** This image of the Sun, taken on 5–6 June 2012 by NASA's Solar Dynamics Observatory, shows the transit of Venus, an event that will not happen again until 2117.

**BELOW LEFT:** The Sun's chromosphere is the source of ultraviolet radiation. It is thought that, in the first few billion years of its life, the Sun was seven times brighter in the ultraviolet.

*Four billion years ago our planet was under siege. Bombarded by the rocky remnants of the Solar System's foundation, our world was a tortured land of barren rock and dust-filled skies.*

The early Earth was not a place that we would recognise as home – 4 billion years ago, our planet was under siege. Bombarded by the rocky remnants of the Solar System's formation, our world was a tortured land of barren rock and dust-filled skies. The days were short, sweeping by in just five hours as the Earth spun frantically on its axis. Each morning this desolate landscape would have been met with the sight of a rising sun very different from the one we see today. Hanging in the sky was a sun in its infancy. If there had been human eyes to view it, it would have appeared only 70 per cent as bright as it is today, and Earth would have been in a kind of perpetual twilight. This raises an interesting question, because there is strong geological evidence that the temperatures on Earth were very similar to those today, and certainly permitted liquid water to exist on the surface. The reason for the relative stability of Earth's climate as the Sun brightened is still a matter of research, although it is thought that a combination of higher concentrations of greenhouse gasses such as $CO_2$ in the atmosphere and, perhaps, less cloud cover resulting in less sunlight being reflected back out into space, kept surface temperatures high.

The relative lack of brightness, however, was deceptive. Beyond the visible and into the UV, the infant Sun was dazzling. This is because the Sun's outer layers were much hotter than they are today, energised by the star's higher spin rate giving rise to intense electromagnetic heating. Hotter surfaces radiate more of their energy in the high-energy, short-wavelength part of the spectrum – in other words, they are brighter in the ultraviolet.

It is thought that the young Sun was seven times brighter in the ultraviolet during the first few billion years of its life. The UV flux at the top of the Earth's atmosphere would have been similar to that experienced by Mercury today, a planet around 100 million km closer to the Sun. The composition of the young Earth's atmosphere is not well known, but it is unlikely that it was able to absorb such high levels of UV radiation. This suggests that it would have been necessary for life to deal with an intense UV onslaught, which may in turn have driven the evolution of pigments at a very early stage in its history. ◉

*The young Sun was seven times brighter during the first few billion years of its life. It would have been necessary for life to deal with an intense UV onslaught, which may in turn have driven the evolution of pigments at a very early stage in its history.*

**BELOW:** A coronal mass ejection blasts off the surface of the Sun in the direction of the Earth, and is deflected by Earth's magnetic field.

**RIGHT:** The aurora borealis (northern lights) occurs when the solar wind – charged particles from the Sun – is drawn by the Earth's magnetic field to the polar regions.

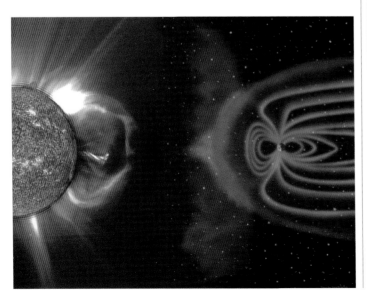

# A UNIQUE AND COLOURFUL WORLD

O n 18 March 2011, after a seven-year journey around our Solar System, NASA's Messenger space probe became the first spacecraft to orbit the tortured inner planet of Mercury. Six days later it reactivated its dormant instrumentation, switching on its powerful cameras and returning the first photograph ever taken from Mercury's orbit. This pioneering spacecraft has sent back thousands of images from the closest planet to the Sun, revealing in extraordinary high definition its complex surface, pitted with craters. But these images also reveal a monochrome world; there is little colour to decorate its dusty, damaged surface.

On its journey to Mercury, the Messenger spacecraft also flew by another of the inner planets – Venus. Again, despite all the acuity of modern technology, this is a planet painted with a limited pallet. A yellow fug shrouds another monochrome planet; a surface with texture but little colour.

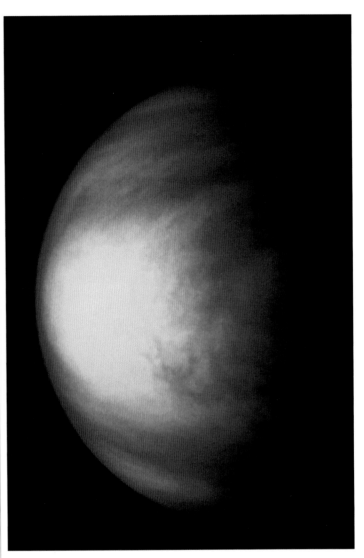

While Messenger was travelling to the inner Solar System, looping around planet after planet to lose enough energy to break its fall towards the Sun, another great adventure was playing out on our sister planet, Mars. Images taken by the now iconic Mars Rovers Spirit and Opportunity have revealed in spectacular detail a planet rich in geology and promise – perhaps even populated by simple, sub-surface life – but limited in shades.

Messenger did photograph one planet that broke the monochromatic mould, however: our very own planet Earth. Side by side, these images of the rocky planets are quite startling in their contrast; it is only our planet that displays a consistent eruption of colour. Ours is a world painted in colour – a rainbow landscape of greens, blues, reds, yellows and violets. Colour, it seems, is a product of life. ◉

**ABOVE LEFT:** Rocks on Mars (left, taken by the Curiosity Rover), compared with rocks on Earth (right). Both images exhibit rounded gravel fragments, suggesting that Mars, like the Earth, once had surface water.

**ABOVE:** A false-colour image of Venus, made by the Galileo Probe, and showing the planet's monochromatic sulphuric-acid cloud formations.

# THE ORIGIN OF LIFE'S COLOURS

Isaac Newton was the first to demonstrate that white light is made up of a multitude of colours when he famously revealed the rainbow hiding in sunlight in 1671 using a simple glass prism. This multicoloured rain illuminates everything on Earth, but why is life so good at selecting only certain colours to reflect into our eyes?

As a particle physicist, I feel I am permitted to think of everything in terms of the interactions between particles. This is a sensible thing to do, since every experiment conducted in the history of science has shown that the elementary building blocks of nature are particles. To be sure, these particles do not behave like little grains of sand or billiard balls; they are quantum particles, and this allows them to exhibit wave-like behaviour. But they are particles nonetheless, and this applies to light as well as electrons, quarks and Higgs bosons.

I will therefore choose to picture the light from the Sun as a rain of particles – an endless stream of photons that rain down on the surface of the Earth after a 150-million km journey from the surface of the Sun. At a subatomic level, when a photon hits something – a leaf, for example – it hits an electron around an atom or molecule and, if the structure of the molecule is just right, the photon will transfer all of its energy to that electron. If the structure of the molecule isn't right, the photon will not be absorbed. In this way, only photons of certain energies interact and are absorbed, and those energies are determined by the structure of the molecules themselves.

As we have already seen, a photon's wavelength is directly related to its colour. So, another way of saying that pigment molecules interact only with photons of particular energies is to say that they absorb only particular colours of light, reflecting the rest away. This is how pigment molecules work – they interact only with photons of particular energies, and therefore absorb only particular colours of light.

There is a dazzling array of pigment molecules in nature, from carotenes that colour a carrot orange, to polyene enolates, a class of red pigments unique to parrots. In some cases the animals and plants produce the pigments themselves, but in many cases they are absorbed into the organism through its diet. If flamingos didn't ingest beta-carotene from blue-green algae in their diet, their trademark pink colour would quickly turn white.

The selective nature of pigment molecules' interactions with photons is the reason for life's rich and varied colour palette. Think about a green leaf. We see

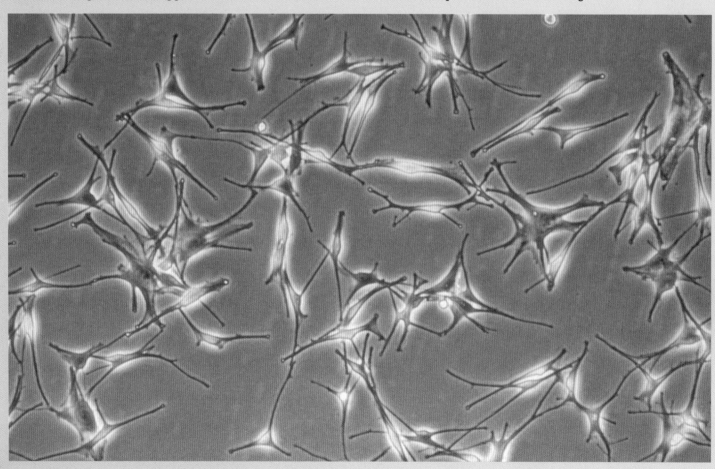

**BELOW LEFT:** A light micrograph of melanocytes (pigment cells), which produce the pigment melanin. It is melanin that absorbs the harmful ultraviolet rays found in sunlight.

**BELOW:** One class of pigments is unique to parrots, and the brightly coloured rainbow lorikeet ('Trichoglossus haematodus') dramatically illustrates life's varied palette.

**BOTTOM:** Flamingos are pink because they ingest beta-carotene from blue-green algae in their diet.

it as green because green photons do not interact with the molecules in the leaf. Red and blue photons do – they are both absorbed by a pigment called chlorophyll. If the rain of photons falls on a surface that reflects the majority of them back (such as the feathers of a swan or the sclera of an eye) we perceive the surface as white. If the light falls on a surface that absorbs photons of all energies (such as a raven's feathers) the surface appears black. The Mexican tiger flower (*Tigridia pavonia*) absorbs all but the lower-energy red photons in sunlight, and so this flower is red. The feathers of the Mexican blue jay (*Aphelocoma wollweberi*) absorb low-energy photons but reflect the higher-energy blue photons back into your eye.

Pigment molecules perform a large variety of functions in living things. Some, as in the case of melanin, evolved to absorb light for protection. It is not known whether the first pigments were used for protection, although many biologists think this was the case. Protection is a simple function, needing no additional complexity such as a nervous system to respond to light's stimulus. There are also pigments that simply make organisms colourful, in order to attract a mate, warn off a predator, entice insects to nectar or invite animals to consume vivid-coloured fruit. But some pigments do not simply dissipate the energy from the Sun as harmless heat or reflect it for display. Chlorophyll, the pigment that lends the natural world its verdant hue, is such a molecule, and its ability to absorb photons and, when integrated into a complex set of molecular machines, use their energy to do something useful, changed the world. ◉

# FROM THE SMALLEST BEGINNINGS...

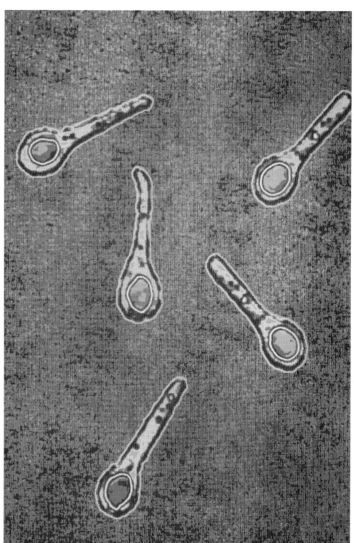

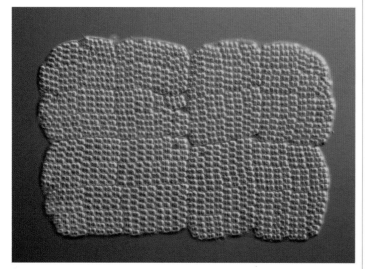

Antonie van Leeuwenhoek was a draper who sold cloth in the Dutch city of Delft in the 1660s. He had no scientific training, was notoriously bad-tempered and could speak only Dutch. And yet this grumpy man literally changed the way we look at the world: he discovered microbes, the most ancient life forms on the planet.

Although the microscope had been invented at the beginning of the seventeenth century, little of much importance had been discovered with it. Then, in Holland, people began to use single-lens microscopes, in which the lens was a tiny ball of glass about 1 millimetre across. Like other Dutch microscopists, such as Jan Swammerdam or the great philosopher Benedict de Spinoza, Van Leeuwenhoek both ground his lenses and pulled them from thin rods of heated glass. But he had an extra trick to produce large quantities of high-quality tiny glass spheres and was so protective of his technique that he kept it secret. What exactly it was is still unknown: his refusal to share his technique may have held back the development of science. This is why, ultimately, the ability to keep secrets is not a common or desirable trait in a scientist – something that legions of conspiracy theorists would do well to remember.

Despite its simplicity, the single-lens microscope gave unparalleled access into the world of the small and revealed amazing wonders – they could magnify up to 500×, revealing

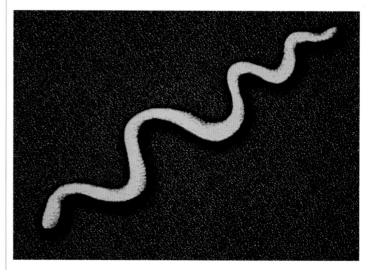

**ABOVE:** Bacteria are organisms known as prokaryotes, which do not have a cell nucleus. The twisted, thread-like spirochaete bacterium shown in this image (here magnified 4,000 times actual size) causes syphilis in humans.

**RIGHT:** Bacteria are found in even the most inhospitable places on Earth. This image shows psychrophilic (cold-loving) bacteria, discovered in Ace Lake, Antarctica, in 1992.

**FAR LEFT:** Van Leeuwenhoek's development of the single-lens microscope led to the possibility of viewing microbial life. This image shows merismopedia – a genus of cyanobacteria whose cells are arranged in perpendicular rows one cell thick to form rectangular colonies.

**LEFT:** The single-lens microscope gave unparalleled access into a previously unseen world. This image is a micrograph of the bacterium *Clostridium tetani*, a rod-shaped, anaerobic bacterium of the genus species Clostridium that causes tetanus.

**BELOW LEFT:** Bacteria have existed for almost the entire history of life on Earth. This computer-generated image depicts a small group of the bacteria *Treponema pallidum*, which causes diseases such as syphilis, bejel, pinta and yaws.

**BELOW:** The single-lens microscope was the forerunner of the scanning electron microscope (SEM), which provides a far greater level of magnification. This image shows *Clostridium botulinum*, the cause of botulism in humans.

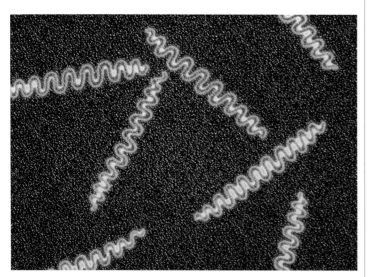

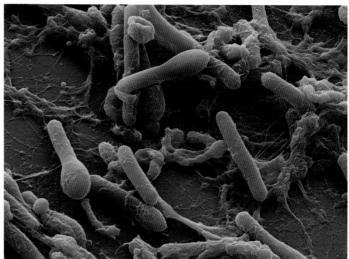

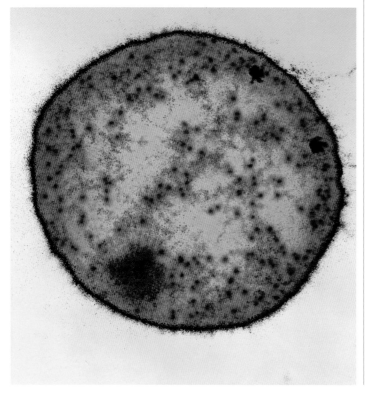

incredible detail in the anatomy of insects and plants. In 1672, Van Leeuwenhoek had been introduced to the Royal Society in London, which invited the Dutchman to turn his powerful device to all sorts of substances, including pepper.

In the spring of 1676, Van Leeuwenhoek tried to discover why pepper was hot; he assumed that there must be some tiny spikes on the surface of the pepper that would explain the tongue-tingling sensation of heat generated by peppercorns. His initial attempts to confirm the 'spiky pepper hypothesis' failed when he used dried pepper, so he put a handful of peppercorns in water and let them soften for three weeks. On 24 April, he drew some of the 'pepper-water' into a glass capillary tube with an extremely fine bore, fixed the tube in front of the metal plate that held his tiny lens in place, and held the apparatus up to the light. To his amazement he saw that the water was full of an incredible number of 'animalcules', or tiny 'animals'. As he wrote to the Royal Society, they 'were incredibly small, nay so small, in my sight, that I judged that even if 100 of these very wee animals lay stretched out one against another, they could not reach to the length of a grain of coarse sand.' These were in fact protists and bacteria. The scale of life had just become almost infinitely smaller than had been imagined.

When the Royal Society got news of Van Leeuwenhoek's astonishing discovery, they instructed their resident microscopist, Robert Hooke, to replicate the 'pepper-water' experiment. He failed, because Van Leeuwenhoek had neglected – perhaps deliberately – to make clear that he had used a capillary tube. Eventually, Hooke realised what was missing from his set-up and was able to confirm the observation. For many years it was thought that the use of pepper infusion was necessary to observe the animalcules – some classes of bacteria are still called infusoria. Of course, the bacteria and protists were in the water all along.

Although he didn't realise it at the time, Van Leeuwenhoek was the first to observe bacteria, the most numerous and ancient life forms on the planet. He went on to explore and detail many uncharted aspects of the biological world – a year later he made the momentous discovery of spermatozoa – but it is rightly for his discovery of the bacteria that he is remembered.

*It is estimated that there are around $10^{31}$ bacteria alive on Earth – a hundred million times the number of stars in the observable Universe.*

Bacteria have been around for almost the entire history of life on Earth. The oldest known bacterial fossils are almost 3.5 billion years old. Typically just a few millionths of a metre in length, these single-cell life forms come in a multitude of forms, from spheres and rods to spirals or even cuboidal shapes. A single drop of water contains, on average, a million bacteria; a gram of soil may be home to 40 million; in your body there are ten times as many bacteria as there are human cells. It is estimated that there are currently around $10^{31}$ bacteria alive on Earth – a hundred million times the number of stars in the observable Universe. By mass, they are comfortably the dominant organisms on our planet.

Bacteria are organisms known as prokaryotes, which means that they do not have a cell nucleus. They share this trait with another group of single-celled organisms known as archaea. The lack of a nucleus, and indeed virtually any complex structures inside their cells, distinguishes them from all other forms of life, which are known as eukaryotes. All animals, plants, fungi and algae – in fact, anything that we would regard as 'complex' – are eukaryotes. The overwhelming majority of biologists today believe that eukaryotes emerged from prokaryotes around 2 billion years ago, and that this fundamental and revolutionary change happened only once. We'll return to this quite remarkable claim later in this chapter. For now, we'll remain in the domain of the prokaryotes, and explore a more ancient yet no less epochal leap in life's capabilities, achieved purely by the seemingly lowly bacteria, that turned the planet green and paved the way for the eukaryotes to flourish. ◉

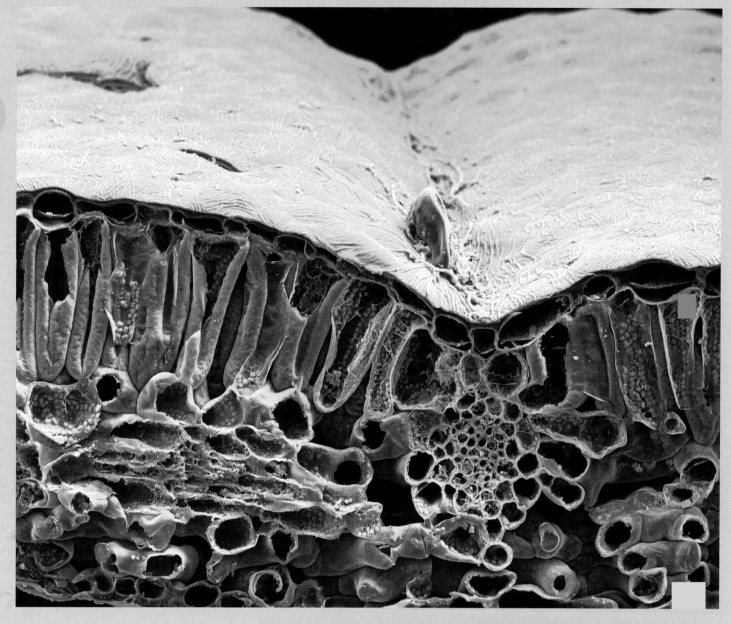

# EATING THE SUN

**ABOVE:** For a plant, one of the purposes of oxygenic photosynthesis is to capture energy from the Sun. This coloured micrograph of the leaf of a Christmas rose ('Helleborus niger') shows the vertical cells of chloroplasts, which perform this function.

If you made it through school biology lessons, you will have heard of photosynthesis. Indeed, you may well be able to recite the famous chemical equation from memory:

$$6CO_2 + 6H_2O \longrightarrow C_6H_{12}O_6 + 6O_2$$
Energy from the Sun

Photosynthesis uses carbon dioxide and water to produce sugars and oxygen in a process powered by the energy of the Sun. But the use of the term photosynthesis to describe this particular process is a colloquialism. Specifically – and this is most definitely not a pedantic distinction – the above equation refers to oxygenic photosynthesis, and this makes all the difference in the world.

Perhaps the best way to unravel the evolutionary origins of photosynthesis, and explain the significance of the term oxygenic, is to look at it from the perspective of a plant. The purpose of photosynthesis, if you are a plant, is twofold. One is clearly visible in the famous equation: it is to make sugars, which is done by forcing electrons onto carbon dioxide. The other, which is hidden in the detail, is to capture energy from the Sun and store it in a usable form. All life on Earth stores energy in the same way, as a molecule called adenosine triphosphate, or ATP. This suggests strongly that ATP is a very ancient 'invention', and the details of its production and function could provide clues as to life's origin 4 billion years ago.

Photosynthesis, therefore, has a dual job: to store energy and to make sugars. The rest of the equation – and in particular the oxygenic bit, which refers to the production of oxygen – is a largely irrelevant detail as far as a plant is concerned. This provides a clue as to how oxygenic photosynthesis evolved.

The molecular machinery of oxygenic photosynthesis in constructed from three distinct components known as photosystem I, photosystem II, and the Oxygen Evolving Complex, linked together by two electron transport chains. This linked molecular machine is known as the Z scheme. Photosystem I takes electrons and, using energy from the Sun collected by the pigment chlorophyll, forces them onto carbon dioxide to make sugars. Photosystem II functions in a different way. It uses another form of chlorophyll and, rather than forcing its energised electrons onto carbon dioxide, it cycles them around a circuit somewhat like a battery, syphoning off a little of the Sun's captured energy and storing it in the form of ATP.

In order to make sugars and ATP, therefore, the plant needs sunlight, carbon dioxide and a supply of electrons. It doesn't 'care' where those electrons come from. The plant may not care, but we certainly do, because plants get their electrons from water, splitting it apart in the process and releasing a waste gas (oxygen) into the atmosphere. This is the source of all the oxygen in the atmosphere of our planet, and so understanding the evolution of the Z scheme is of paramount importance if we are to understand how Earth came to be a home for complex animals like us. The story can be traced back over 3 billion years to a time when the only life on Earth were the single-celled bacteria and archaea. ◉

## PHOTOSYNTHESIS

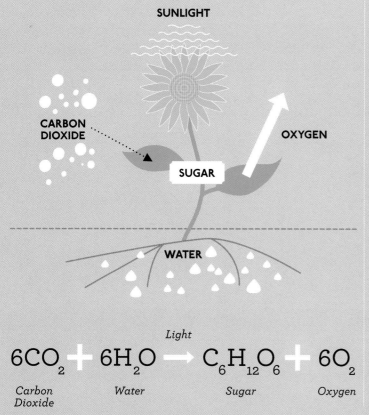

$$6CO_2 + 6H_2O \xrightarrow{Light} C_6H_{12}O_6 + 6O_2$$

Carbon Dioxide     Water     Sugar     Oxygen

## CONVERSION OF WATER TO OXYGEN AND LIGHT TO ENERGY

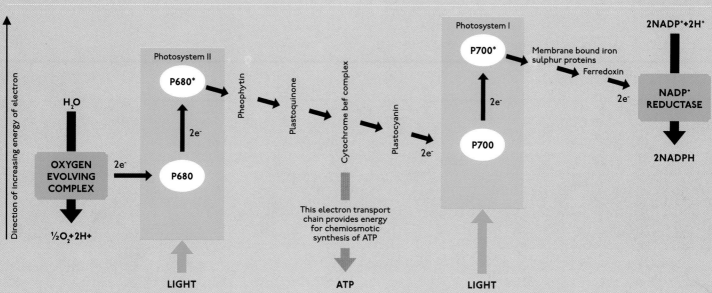

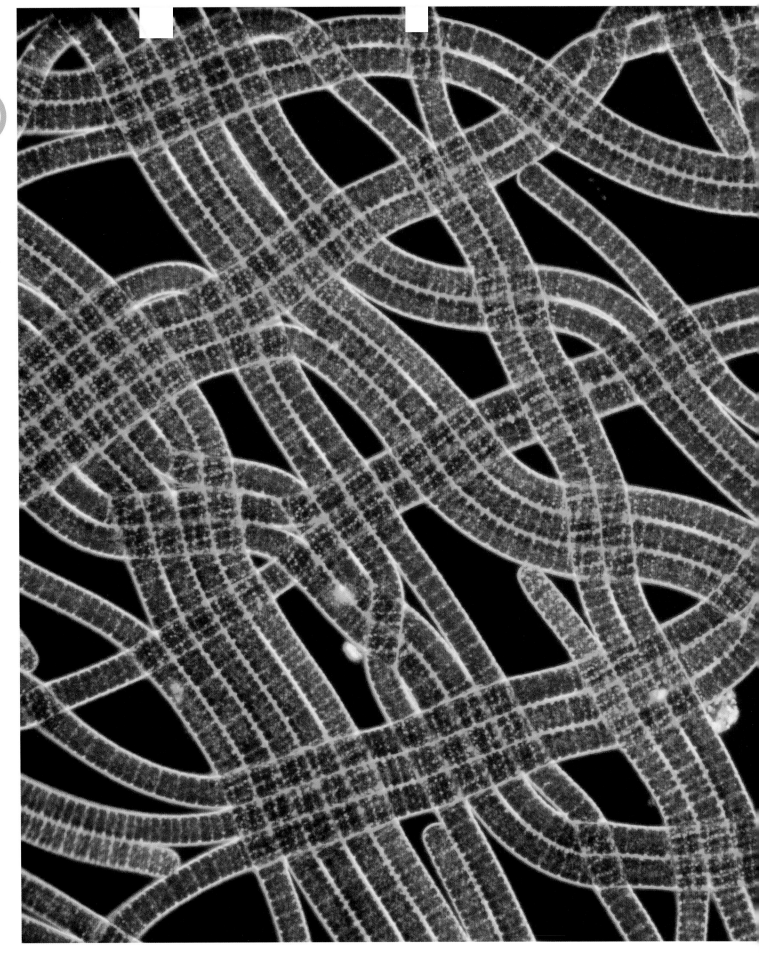

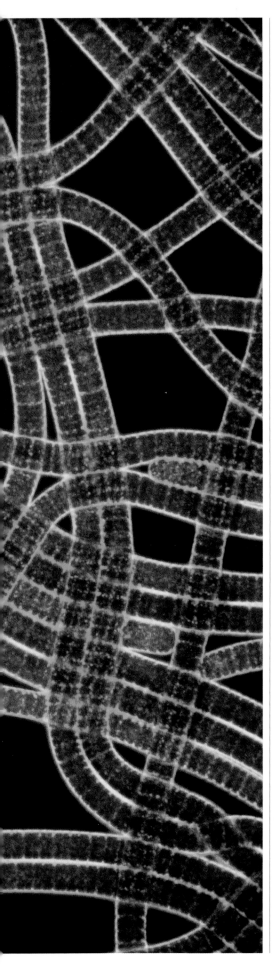

*Take a look at this picture – it's an image of a very particular type of bacteria. Look very closely at it because you have a lot to thank this particular kind of organism for. These are cyanobacteria – lowly bacteria that sit at the very bottom of the food chain. They're the most numerous organisms on the planet. There are more of them on Earth than there are observable stars in the Universe, and these little creatures are what enabled you – and every other complex living thing that has ever lived on the planet, from dinosaurs to daffodils – to exist.*

*If you look at the picture carefully, you will see that, unlike the other monochromatic bacteria, this one is bursting with a kind of blue-green colour, which comes from a pigment known as phycocyanin – exactly the kind of pigment that would offer an organism protection from the Sun's damaging UV radiation. But these bacteria don't just use the pigment for protection, they use it to capture the energy of the Sun.*

**LEFT:** This light micrograph shows cyanobacteria, or 'blue-green algae', which use phycocyanin to capture the energy of the Sun.

# A BREATH OF FRESH AIR

**ABOVE:** Cyanobacteria are able to reproduce rapidly, and this can have a devastating impact on an ecosystem. This satellite image of Lake Atitlán in Guatemala shows blooms of cyanobacteria, caused by polluted runoff from the surrounding land.

Today cyanobacteria are sometimes considered to be a problem. This image, although beautiful, is of a bloom of 'blue-green algae'– or, more correctly, cyanobacteria – in Lake Atitlán in the Guatemalan Highlands. It provides a vivid example of bacteria reproducing at a ferocious rate, and, in some cases, this explosion of life can have a devastating effect on an ecosystem. Toxins produced by the bacteria can decimate water life and affect human health, so they are closely monitored by environmental agencies around the world. But we have cyanobacteria to thank for the oxygen we breathe, because it is a virtual certainty that oxygenic photosynthesis evolved in an ancient cyanobacterium.

The way to unravel the story of the evolution of the Z scheme is to look at how each individual part may have arisen. There is evidence that an early form of photosynthesis may have emerged as far back as 3.5 billion years ago in single-celled organisms that produced enigmatic mounds known as stromatolites (see Chapter 3), although the precise date is still an area of active debate and research. Whatever the date, there is general agreement that a simple form of photosynthesis, using energy from the Sun to synthesise sugars from carbon dioxide, just as photosystem I does in plants today, is very ancient. The pigment used today is chlorophyll, a member of a family of molecules known as porphyrins. Complex though they are, porphyrins have been found on asteroids, implying that they form naturally and are likely to have been around on Earth before the origin of life. There are still bacteria alive today that have only photosystem I. They take their electrons from easy targets, such as hydrogen sulphide or iron, and don't therefore need much else in the way of machinery.

Over time, it is thought that some bacteria adapted this early photosynthetic machinery to perform a different task – the production of ATP. There are similarities between the two photosystems that strongly suggest a common origin and later specialisation.

evolutionary adaptation now presents itself. What if it were possible to run these two machines at once, connecting the electron circuit from photosystem II into photosystem I, which would dutifully dispose of the cascade of electrons by pushing them onto carbon dioxide to form sugar? This would confer a great advantage on the organism in question, allowing it to make both food and ATP at the same time using sunlight as an energy source. This is certainly a plausible explanation for the separate evolution and then recombination of the two photosystems, but it leaves one remaining question: where does this machinery get its electrons? Here is where the Oxygen Evolving Complex enters the story and, with it, one of the most important evolutionary steps in the history of life on Earth.

The Oxygen Evolving Complex is an odd structure: more mineral than biological. It consists of four manganese atoms and a single calcium atom, held together in a lattice of oxygen. Manganese is locked away in vast mineral deposits on the ocean floor today, but in the early history of our oceans it would have been available in seawater for organisms to use. Bacteria use manganese to protect them from UV light, in much the same way as we use melanin – manganese is easily 'photo-oxidised', absorbing the potentially harmful UV photon and releasing an electron in the process. This

*Bacteria genes can be switched on and off, allowing them to make hay while the Sun shines – or at least to use sunshine to make sugar or ATP.*

The evolution of early versions of photosystems I and II in bacteria is therefore relatively well understood; their components are simple, and the chemistry reflects that occurring naturally on the early Earth. Things become more interesting, however, when we ask how these two machines came to be joined together in the Z scheme. While biologists don't yet agree on the answer, one of the more elegant hypotheses, due to Professor John Allen at Queen Mary, University of London, and detailed in Nick Lane's excellent book, *Life Ascending*, is as follows.

While some bacteria employed the precursor of photosystem I, and others used the precursor of photosystem II, there may also have been bacteria that possessed the genetic coding necessary to build both photosystems. This would allow them to switch between them, depending on environmental conditions and the availability of food. This is a relatively common thing for bacteria to do today; their genes can be switched on and off, allowing them to make hay while the sun shines – or at least, in this case, to use sunshine to make sugar or ATP, depending on whether the imperative is to reproduce or simply to survive. The possibility of an ingenious

may have been one of the ways in which electrons made their way into the primitive photosystem II in early bacteria. So manganese, at least, was already an important component of living things from the earliest of times. Today, manganese performs a different task. It sits at the heart of the Oxygen Evolving Complex, whose job is to grab water molecules and hold them ready for electrons to be ripped off and used as input into photosystem II. As a result, water molecules are split apart and, just as in the electrolysis of water so beloved of Mr Bell (see page 27), oxygen is released as a gas.

This theory is a piece of cutting-edge research. The structure of the Oxygen Evolving Complex was determined only in 2006, and it is only in the last few years that the locations of each of the 46,630 atoms in photosystem II have been mapped. There are therefore many details in this story yet to be uncovered, but the broad sweep we have outlined here is certainly a strong candidate for an explanation of how the complexity of the Z scheme arose.

There is one last quite wonderful sting in the tail of this story, however, and it is something we know for certain: oxygenic photosynthesis evolved only once.

The evidence for this rather definite statement is clear when we look down a microscope at the structures inside plants and algae that carry our photosynthesis. They are called chloroplasts, and they are all self-evidently related to each other because they are so similar. But there is more than this, because they look for all the world as though they were cyanobacteria living inside the leaves, just like those found today in the blooms on Lake Atitlán. This is because that is exactly what they are. They even maintain their own independent rings of DNA, just as free-living bacteria do today.

But how does one cell end up inside another? At some point in the history of life on Earth, a cyanobacterium cell must have been engulfed by another cell and, instead of being digested, it survived to perform a useful purpose. This process, called endosymbiosis, has happened more than once in the history of life on Earth; indeed, it is thought to have been fundamental in the evolution of complex life.

**BELOW LEFT:** A coloured electron micrograph of a leaf of *Zinnia elegans*, showing chloroplasts (green), starch granules (pink), the nucleus (red), and a large vacuole (white). The large air spaces allow for gas exchange during photosynthesis.

**BELOW:** This coloured electron micrograph shows two chloroplasts in the leaf of a pea plant (*Pisum sativum*). Chloroplasts convert light and carbon dioxide into carbohydrates.

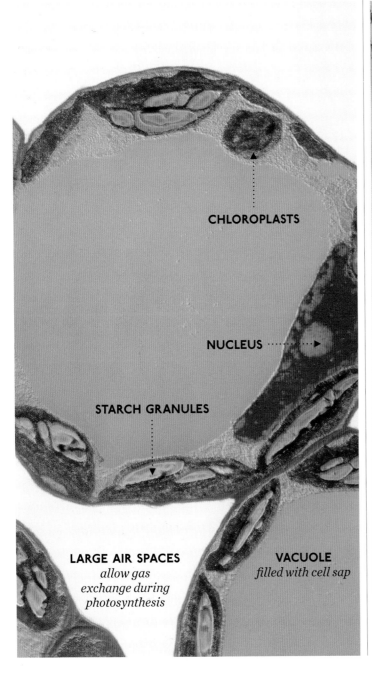

**CHLOROPLASTS**

**NUCLEUS**

**STARCH GRANULES**

**LARGE AIR SPACES**
*allow gas exchange during photosynthesis*

**VACUOLE**
*filled with cell sap*

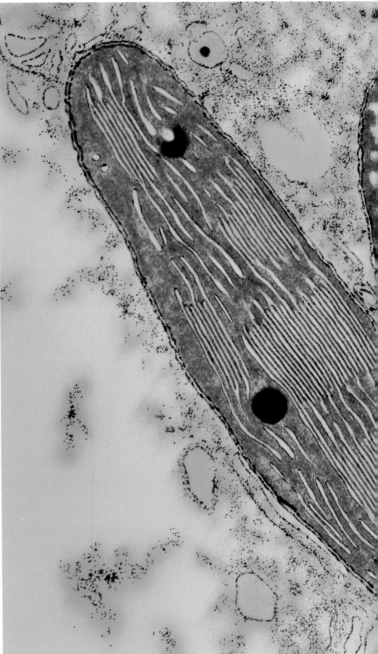

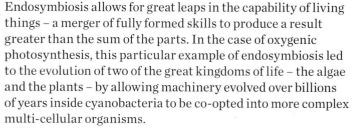

**BELOW RIGHT:** A coloured electron micrograph of the inside of a chloroplast's thylakoid membrane, containing the green pigment chlorophyll.

Endosymbiosis allows for great leaps in the capability of living things – a merger of fully formed skills to produce a result greater than the sum of the parts. In the case of oxygenic photosynthesis, this particular example of endosymbiosis led to the evolution of two of the great kingdoms of life – the algae and the plants – by allowing machinery evolved over billions of years inside cyanobacteria to be co-opted into more complex multi-cellular organisms.

The quite dizzying conclusion is that, because everything that carries out oxygenic photosynthesis today does so in precisely the same way, we owe the beauty of life on Earth – with its hues, colours and seemingly limitless diversity – to a cyanobacterium whose ancestors, somehow, found their way inside another cell. The descendants of that cell are still present on Earth today, inside every leaf, every blade of grass and every algal bloom, and they have filled our atmosphere with oxygen. ◉

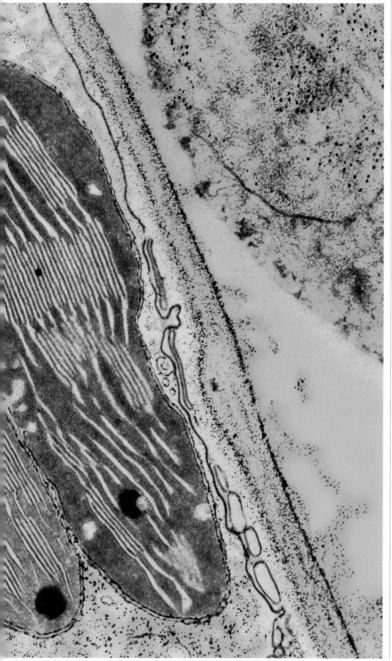

# BREATH OF LIFE

For almost half of Earth's history, one of the most important ingredients for complex life was absent from the Earth's atmosphere. Oxygen is an unstable, reactive gas that must be constantly replenished. The first rush of oxygen released from water by the cyanobacteria did what oxygen does best, reacting with the myriad elements present on Earth's primordial surface to form oxides. In our planet's infancy, large amounts of iron could be found in the oceans and, to a lesser extent, on land. Left over from the Earth's formation, this dissolved iron remained stable for billions of years, but as the levels of atmospheric oxygen began to rise, a very familiar reaction began to take place. The Earth began to rust. Today, across the planet the evidence of this global rusting can be found in deposits of iron oxides known as banded iron formations.

Oxygenic photosynthesis doesn't automatically fill the atmosphere with oxygen, however. It is necessary, but not sufficient, because both rusting and respiration act to undo all the good works of the plants, algae and cyanobacteria, and remove oxygen from the atmosphere. While photosynthesis

**BELOW:** As levels of atmospheric oxygen rose, the Earth began to rust. Evidence of this rusting can be seen at Rockham beach, North Devon, where deposits of iron oxide appear as orange patches.

takes carbon dioxide out of the atmosphere and turns it into organic matter, aerobic respiration takes organic matter and burns it using oxygen, releasing carbon dioxide and water. These processes will naturally reach a balance, which is why oxygen levels today have been stable at around 21 per cent for many millions of years. In order to change oxygen levels, something has to happen. It is known that oxygen levels first increased on Earth around 2.4 billion years ago, a time when many of the great banded iron formations were laid down. This rise may have been triggered by the complete oxidation of the Earth's iron and other elements, which until that time acted as a sink, removing the photosynthetic oxygen from the atmosphere as quickly as the bacteria could release it. This is one plausible scenario, although there is not widespread agreement on the reason for this 'Great Oxidation Event', and it is still a very active area of research. Whatever the reason, the oxygenation of the atmosphere made possible by the evolution of oxygenic photosynthesis was critically important for the emergence of complex animals.

Aerobic respiration, in which energy is released from organic matter, makes the existence of food chains possible because it is so efficient. Releasing energy from food using oxygen is around 40 per cent efficient, while oxidising food using iron or sulphur is only around 10 per cent efficient. This means that animals can eat plants, and in turn get eaten by a tower of predators that can still extract enough energy to flourish. It is almost certainly no accident that the Cambrian explosion – the rapid diversification of life resulting in the emergence of virtually everything we would regard today as complex – followed (on geological timescales) a rapid increase in atmospheric oxygen levels.

## BRINGING IT ALL BACK HOME

The story of the emergence of today's Earth is complex. That we understand not only the broad sweep of the narrative, but also the fine detail of at least some of the chapters, is one of the great achievements of science, and the presence of some uncertainties in the story of the emergence and development of a 4-billion-year-old biosphere is surely unsurprising. We have seen that water is a prerequisite for life on Earth, and most likely for life anywhere in the Universe. Likewise, an oxygen atmosphere, while not necessary for microbes, is a vital component of a complex ecosystem able to support large predators and prey, and probably therefore intelligent civilisations. As oxygen atmospheres are inherently unstable, oxygenic photosynthesis on a global scale is necessary to maintain high levels of this life-giving gas. And we know that this evolved only once on Earth. But there is one final ingredient that is more elusive and certainly beyond life's control: time. It is a certainty that the evolution of complex life requires an ecosystem that is stable over many millions of years. But how many millions? This question will occur again and again throughout this book. Why did life emerge so soon after the birth of our planet, only half a billion years after its formation? And how did the first life blossom into the magnificent complexity we see on Earth today? A good place to start is to look at the evolutionary history of a single animal, and see how precisely we can trace its origins back into the deep past. ◉

**VARIATION OF ATMOSPHERIC OXYGEN CONCENTRATION OVER THE LAST 3.5 BILLION YEARS**

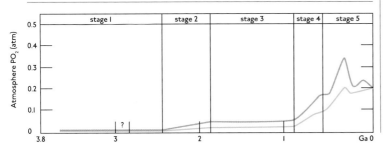

# FOUR-LEGGED LIFE STORY

**EXMOOR PONIES**
The Exmoor pony, an example of the modern horse ('Equus caballus'), and little changed since the last ice age.

The family tree of the horse is the best known of any complex animal, partly due to the wealth of available fossil evidence. The first animal recognisably 'horse-like' is the Hyracotherium (once known as 'Eohippus' or 'dawn-horse'), which lived around 50 million years ago. It was a fox-sized omnivore, and because many thousands of intact skeletons have been found, a great deal is known about its form and lifestyle. Alterations in the availability of food probably played a role in the later emergence of two other species, the Orohippus and Epihippus, both better adapted to a browsing diet of tough plants. Around 30 million years ago, changes in climate saw the emergence of grasslands and steppe landscapes across the planet. In North America, the Mesohippus emerged; with longer legs and a slightly larger frame, it was better adapted for life on the new grassy plains because it could run faster to avoid predators. At around the same time, a species known as the Miohippus appears in the fossil record. It probably lived alongside the Mesohippus, but over time gradually replaced it. This raises an important point relating to the construction of a family tree. It should not be read as a gradual transition from the simple to the complex, culminating somehow in the grandeur that is a modern domestic horse. The different species were adapted to different conditions, occupying different ecological niches. None was 'better' than another in any absolute terms. Because all that remains of these animals are their fossils, and our knowledge of the local ecology is generally rudimentary, it is often difficult to know why one species survived while another died out, or why a particular species arose through a process of 'speciation'. We will explore the phenomenon of speciation in much more detail in Chapter 5. The domestic horse should therefore be seen as one of many branches of a complex family tree that survives today – it is not the culmination of a series of 'improvements' over its more distant ancestors.

That said, it is still instructive to follow the tree, because it illustrates the surprising pace of change delivered by the power of evolution through natural selection. Over around 25 million years, the Miohippus has given rise to a grand array of species, forever passing on a shifting mixture of genes, filtered by the sieve of natural selection. As conditions changed, some branches of the tree turned out to be evolutionary cul-de-sacs, while others lived on, branching or gradually changing. This continuously shifting selection and isolation of pools of genes is reflected in the diversity of the horse family that we can see today – from zebra to the Yucon wild ass, from the kiangs of Tibet to *Equus ferus caballus* (the domestic horse).

**1 MILLION YEARS AGO**

**1.6 METRES**
Modern horse (Equus)

**1.25 METRES**
Pliohippus

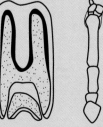

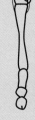

Molar teeth   Forefeet

## MERYCHIPPUS
Artist's impression of Merychippus, much smaller than the modern horse, and with three toes on each foot.

## HYRACOTHERIUM
Artist's impression of Hyracotherium, which lived around 50 million years ago, and which is thought to be the first true horse.

15 MILLION YEARS AGO

100 MILLION YEARS AGO

**1.0 METRES**
Merychippus

**0.6 METRES**
Mesohippus

**0.4 METRES**
Hyracotherium

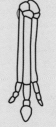

So we can see that changes in form can be rapid and surprising. The Hyracotherium looks more like a fox, or even a large rodent, than a horse, and yet in the history of every modern horse, Przewalski's horse, zebra or donkey there are Hyracotherium ancestors who lived only 50 million years ago – the blink of an eye in the 4.5 billion years' life of our planet.

If we sweep back still further, we encounter the first mammals around 225 million years ago. There is an explosion of complexity in the fossil record associated with the Cambrian period, 530 million years ago, which may have been related to a rise in oxygen levels. The first evidence of complex multicellular life appears around 600 million years ago in the form of the Ediacaran biota, named after fossils first found in the Ediacaran hills in Australia. Some of these organisms had a quilted, fractal appearance and were so bizarre that it has been suggested they were neither animals, nor plants, nor fungi, but some failed evolutionary experiment. Other Ediacarans were clearly soft-bodied animals, up to 2 cm in length with a head, and may even have burrowed slightly into the microbial mats on the sea bottom, thereby subtly changing the planet's ecology, and opening the way for further evolutionary developments. The delicate and ambiguous nature of the fossils left by these mysterious organisms has made the study of the Ediacara one of the most intriguing parts of recent palaeontology.

Before the earliest Ediacaran fossils, dated at 655 million years old, there is no direct evidence of multicellular life on Earth. The next major milestone occurred around 2 billion years ago with the emergence of the eukaryotic cell. As we have already discussed, eukaryotes are cells with a nucleus and internal structures similar to our own – we are grand colonies of eukaryotic cells. At around 3.5 billion years ago, we find the first prokaryotes, the first free-living cells that emerged, perhaps, from hydrothermal vent systems on the ocean floor of the primordial Earth.

The reasons for these vast periods of apparent stasis in the development of life – over a billion and a half years from prokaryote to eukaryote and a similar amount of time from the eukaryote to the first evidence of multicellular life – are not understood. It certainly seems that the complex cell – the eukaryote – emerged only once in the history of life. There is no evidence of different versions of the eukaryotic cell emerging from bacteria or archaea during their 4 billion-year tenure on Earth. If they did, they have left no trace. All animals, plants, algae and fungi are self-evidently related to each other, sharing multiple traits from the structure of their DNA to the use of ATP. The form and biochemistry of their cells are very similar; only the colonies they form are radically different. This strongly suggests a common ancestor, and raises the intriguing question: was the emergence of the eukaryote an incredibly unlikely event, or was the billion-year delay just bad luck? We don't know, because we have only one Earth to observe. This is why the search for microbial life elsewhere in the Solar System is so desperately important.

There is something on which everybody agrees, however: as first proposed by the late Lynn Margulis, the eukaryote is a chimera, formed by endosymbiosis in much the same way as the ancestors of plants and algae acquired their chloroplasts. The evidence for this lies in structures known as mitochondria, found in the overwhelming majority of eukaryotic cells alive today and responsible for the generation of ATP through respiration. We will meet mitochondria in more detail in Chapter 2. Just as for the case of the chloroplasts, however, mitochondria have their own loops of bacterial DNA that mark them out as ancient bacterial symbionts. Furthermore, eukaryotes bear an interesting genetic relationship with the two prokaryotic branches of life – bacteria and archaea. They share genes with both, which strongly suggests that the first eukaryote was the result of a merger between a bacterium and an archaea. The details are still the subject of debate, but it seems that something very unlikely indeed – the successful merger of two prokaryotic cells – had to happen before complex life could develop on our planet. The eukaryote is probably a happy accident. And, therefore, so are we.

This may have profound implications for the existence of complex life on other planets. The emergence of prokaryotes may well be inevitable, given the right conditions. We will explore this in much more detail in the following chapter. But there is no way we know of that prokaryotes will clump together to form animals, plants and people – at least they haven't managed it during their 4 billion years on Earth. To build Apollo 8, you first need a eukaryote, and it seems probable that on Earth this key step occurred due to blind luck, followed by a lot of natural selection. It took almost 2 billion years for it to happen on our planet – 2 billion years in which the oceans remained a stable, hospitable home under a dangerous Sun. Could it be that living things capable of taking pictures of their home from space are rare or even unique to Earth? Perhaps one day we will know, but, lonely or not, we have surely learnt enough about life's long struggle to complexity to treasure the good Earth. ◉

**RIGHT:** 'Dickinsonia costata' – discovered in Ediacara, Australia – is an iconic fossil of the Ediacaran biota, dating from around 600 million years ago, when the first evidence of complex multicellular life appears.

# CHAPTER 2

—

# WHAT IS LIFE?

# THE BRIEFEST OF BEAUTY

For as long as science has explored the great mysteries of the Universe, it has wrestled with the idea of a life force, a spirit or a soul. For over two thousand years the greatest minds on the planet have attempted to unlock the secret force that appears to divide the living from the dead. From Aristotle and the ancient Greek philosophers, to the great thinkers of the modern era such as Descartes and Kant, the search for the essence of life has been a long and often futile pursuit better left to religion than rationalism. Even with the advent of modern science, attempts to pinpoint a physical basis for the dividing line between life and death have often struggled; Phlogiston theory and Vitalism are just two of the defunct notions that held sway over scientific thought in centuries gone by. The last 100 years, however, have taken us tantalisingly close to being able to explain the basis of life in the most physical of terms.

**RIGHT:** Night falls on the Day of the Dead festival in Sagada, the Philippines: this is a time of celebration rather than bereavement.

# DAY OF THE DEAD

The town of Sagada lies 320 km miles north of Manila in the mountainous north of the Philippines. The journey takes two days by road, most of which – in my memory at least – consists of being battered and rattled inside a Jeepney. These flamboyant contraptions are heavily modified World War II Jeeps, left by the departing American army, on which has been conveyed a supernatural longevity defying all known engineering principles. We may one day be able to explain life, but the mechanical survival of the Jeepney is truly miraculous.

The history of the north is quite different from that of the rest of the country, largely because of its inaccessibility. A Spanish mission was established in the late 1800s, bringing with it a veneer of Christianity that overlays much older tribal beliefs (although few Conquistadors made it this far without Jeepneys). This relatively light colonial touch has resulted in a unique and powerful take on a celebration known as the Day of the Dead.

In modern times, this festival has become associated with the more familiar Christian celebrations of All Saints' Day, and the probably pre-Christian Halloween, but its routes can be traced back to the pre-colonial indigenous Mexican civilisations. Cross-cultural pollination can appear surprisingly random! Sagada's take on this Aztec festival, delivered by Catholic missionaries, is further seasoned by the local tribal belief in animism, visibly displayed in the town's famous hanging coffins. Animism holds that a life force or soul is not unique to humans, but exists in all things, from the lower animals to trees, lakes and mountains. By returning the dead to this picturesque rock face, it was thought that the spirits could return to the living mountain.

This potent brew of superstition is realised in spectacular and moving fashion on the evening of 30 October, when the villagers of Sagada gather in a quite magical hillside churchyard to spend an evening with their deceased loved ones. As night falls and the white gravestones shine in the subtropical winter light, each family lights a fire beside their ancestral grave and the landscape shifts instantly from English country idyll to post-apocalyptic film set. The intensity of the heat and smoke is something to behold, and acts as a crackling orange backdrop to a genuine sense of celebration.

**ABOVE:** The hanging coffins of the limestone rock face in Echo Valley, near Sagada. In animist belief systems, the souls of the dead return to the living mountain.

This took me by surprise; the Day of the Dead suggested something more sombre to me, but the feeling on the hillside was one of family reunion rather than bereavement. 'I believe that the spirits are around us,' said one woman. 'I know that they are near.' No matter how unscientific this sounds, it is inappropriate to dismiss her statement, or the beliefs of the people on the hill, without thought. It certainly feels right. Or at least it feels right in the sense that the alternative feels wrong. I don't believe it, which means that when I die, I will be nothing more than an inanimate bag of chemicals slumped on the floor. Nothing has left, yet what is left is no longer me. That sounds almost contradictory. If it is not to be so, then it is incumbent on science to explain how the feeling I have of being alive, and all the processes associated with living, emerge from a simple collection of chemicals that can be found anywhere in the Universe. This is a tall order, and while we don't yet have all the answers, we have made significant progress towards answers that are certainly plausible. ◉

# WHAT IS LIFE?

*The modern scientific view is that life began, if not in Promethean clay, then in wet rocky chambers, and there is an unbroken line that leads, over 4 billion years, to us.*

**LEFT:** Erwin Schrödinger understood the human equivalent of a descent into chaos only too well when he wrote *What is Life?* Just a few years earlier, he had been caught in the unfolding political turmoil and violence of Europe, becoming one of the most high-profile scientists to flee after the Nazi invasion of his Austrian homeland. Leaving Graz in September 1938, Schrödinger left behind not only his academic position but even his Nobel medallion as he and his wife Anne smuggled themselves out of the country with just a handful of belongings. Via Rome and then Oxford, Schrödinger would finally settle in Dublin in 1940, on a personal invitation from the Irish government to help set up the Institute for Advanced Studies. It was in Dublin in 1943 that he gave the public lecture series that became *What is Life?*

In February 1943, the physicist Erwin Schrödinger gave a series of lectures at Trinity College, Dublin, entitled 'What is Life?' Schrödinger is undoubtedly best known for his seminal work on Quantum Theory, for which he shared the 1933 Nobel Prize in Physics with Paul Dirac. In his lectures, and the subsequent book of the same title, he addressed a problem undoubtedly more difficult and challenging than his famously counterintuitive description of the subatomic world. 'How can the events in space and time,' asked Schrödinger, 'which take place within the spatial boundary of a living organism, be accounted for by physics and chemistry?' Within a single paragraph he provides, if not a partial answer, then at least a statement of intent: 'The obvious inability of present-day physics and chemistry to account for such events is no reason at all for doubting that they can be accounted for by those sciences.' This was – and perhaps still is – a bold assertion, albeit one with which I profoundly agree. Schrödinger argues that life is a physical process, described by the same laws of physics as the falling of the rain or the shining of the stars. How could it not be so? The alternative is to propose the existence of phenomena in the Universe that are specific to life, and that mark out a fundamental difference between the living and the dead. While these phenomena may be amenable to scientific study, they would not be discernable from the study of inanimate objects. This would imply in turn that questions surrounding the origin of life would always remain beyond science, because life by definition possesses something that no inanimate object possesses – for the sake of argument, let's call it a soul. There is of course a rich historical narrative surrounding such speculations. During the late eighteenth and nineteenth centuries, rapid advances in surgical techniques led to commensurate leaps in the understanding of the machinery of the human body. The experimentation on and dissection of dead bodies became a spectator sport, culminating in Giovanni Aldini's spectacular attempts in January 1803 to reanimate the corpse of a murderer, George Foster, by connecting his body to a large number of the newly invented voltaic batteries. Unsurprisingly, the body jolted, waving its hand aloft and, it is said, opening an eye. The popular idea that electricity might be the vital force separating the living from the dead was famously dramatised by Mary Shelley in her novel *Frankenstein; or, The Modern Prometheus*. Central to the novel is the question of whether a re-animated machine, composed of inanimate organic pieces, can ever develop, or perhaps be inhabited by, human morals and feelings. This is a good question because, as we shall see, the modern scientific view is that life began, if not in Promethean clay, then in wet rocky chambers, and there is an unbroken line, characterised by the assembly of an increasing number of organic components, that leads, over 4 billion years, to us.

The debate reached its apotheosis with the publication of Darwin's *On the Origin of Species* in 1859, which removed the artificial distinction between humans and other forms of life for good (both meanings of the word are appropriate). But if we strip away the supernatural, then we are still left with Schrödinger's question, hanging as a challenge to which science must rise if it is to provide a complete description of the Universe: What is Life? ◉

# ENERGY AND THE FIRST LAW OF THERMODYNAMICS

I n 1840 a young German doctor called Julius von Mayer was hired as a ship's physician aboard a Dutch merchant ship sailing for Java. As a recently qualified medic, von Mayer would surely have approached the voyage with a mixture of excitement and trepidation, and there would be no shortage of patients to treat with his newly acquired skills of bloodletting. Yet despite his medical duties, it was not the drama of life and death at sea that provided the most profound experience of his travels. Von Mayer fell into conversation with a navigator who shared with him a counterintuitive detail about the ocean: it is warmer when whipped by a storm and cooler in a calm sea.

This observation intrigued von Mayer and, on returning to Germany, he set aside his medical studies to pursue a new-found interest in the physical sciences. With little training in mathematics or physics, he published his first paper shortly after his return in 1841. His paper, which was rejected by the famous journal *Annalen der Physik*, was entitled 'On the Quantitative and Qualitative Determination of Forces'. In it, he stated that 'motion is converted into heat', a first, partial step towards the discovery of one of the fundamental laws of physics: the first law of thermodynamics. The reason that the waves are warmer during a storm, proposed von Mayer, was the same reason that the waves roared as they crashed and the ship rolled and swayed with their impact. In modern terminology, energy is conserved; it is neither created nor destroyed. Waves carry energy, and if they hit a ship they can lose some of that energy, which must be transformed into the motion of the ship, sound or heat.

Von Mayer went on to publish a series of papers investigating the links between heat, energy and work, including the observation that oxidation, as it is now known, is the process by which animals extract energy from food. But because of his lack of contacts and formal education, his work was largely ignored. Credit for the precision experimental work leading to the first law is usually given to the Manchester-based scientist James Joule, which may not be entirely erroneous. Joule was able to convince a sceptical scientific establishment by the sheer precision of his measurements; he famously determined that it takes 772.55 foot-pounds force of work to raise the temperature of one pound of water by one degree Fahrenheit, a number carved into his tombstone in Brooklands cemetery near Manchester. The reason for the arcane units – which will no doubt send

shivers of disapproval through the modern-day unit police, who prefer their physics to be metric – is that, to heat the water, Joule used a falling weight, the mass of which he measured in pounds, and the distance through which it fell he measured in feet. Joule's assertion was that all the energy of the falling weight was converted into heat. Again, energy can be neither created nor destroyed – it is merely converted from one form to another.

Arguments over precedence (who did what first) are common and often vitriolic in science. Many feel that von Mayer deserves credit equal to Joule, although in fact the American (in the sense that he was born in Massachusetts in 1753, before the US Declaration of Independence) Benjamin Thompson – later Count Rumford of Bavaria – made similar

observations at the turn of the nineteenth century while boring cannon for the Duke of Bavaria. So it is fair to say that the conservation of energy was 'in the air' throughout the first half of the nineteenth century, and many of the great scientific names worked in this new field known as thermodynamics.

The reason it took a great deal of experimental work and theoretical wrangling to emerge with the seemingly simple statement of the first law of thermodynamics – that energy is neither created nor destroyed – is that it is counterintuitive. It is not obvious that heat is a physical manifestation of something also possessed by a falling weight. Does the unpleasant feeling you get when you are burned by a hot pan have the same underlying physical description as the unpleasant feeling you get when someone hits you in the face with a cold pan? The answer is yes; in both cases energy from the pan is transferred into your body, and the consequences are unpleasant.

The concept of energy is absolutely central to the description of any physical process, because it is always conserved. Because it can be neither created nor destroyed, all that can happen to it at the most fundamental level is that it is transformed from one form to another. In a sense, this is all there is to the Universe! If no energy is 'flowing' – a colloquialism by which we mean 'being transformed from one form to another' – then nothing is happening at all. Here is the first step on the road to answering Schrödinger's question – What is Life? Whatever it is, it is a process by which energy is transformed from one form to another. ◉

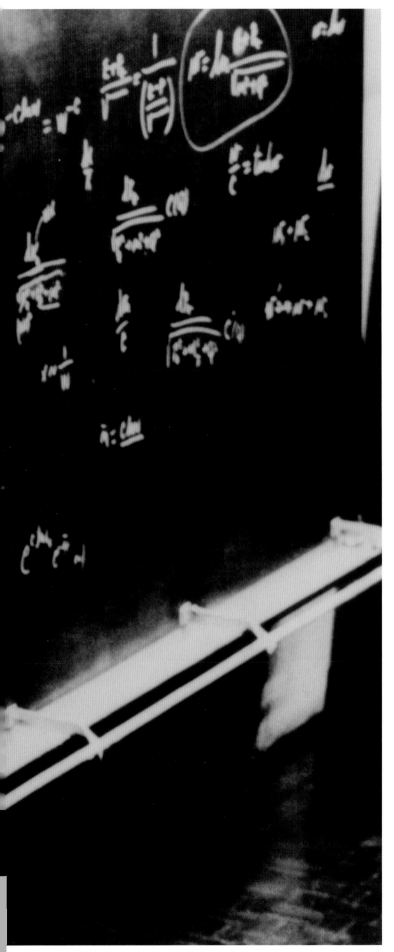

*There is a fact, or, if you wish, a law, governing all natural phenomena that are known to date. There is no known exception to this law – it is exact so far as we know. The law is called the conservation of energy. It states that there is a certain quantity, which we call energy, that does not change in manifold changes which nature undergoes. That is a most abstract idea, because it is a mathematical principle; it says that there is a numerical quantity which does not change when something happens. It is not a description of a mechanism, or anything concrete; it is just a strange fact that we can calculate some number and when we finish watching nature go through her tricks and calculate the number again, it is the same.*

*The Feynman Lectures on Physics*

## FIRST LIFE

The largest island in the Philippines is Luzon. Home to the country's capital, Manila, it is the economic and political heart of the country. However, this is an island that hides much of its power in the extraordinary landscapes that surround the capital. Luzon is covered in mountains and shaped by volcanic activity. Contained within this single island is a geologist's paradise – from the ferocious Mount Pinatubo, one of the most violent volcanoes of the twentieth century, to the beautiful Mayon volcano, perhaps the most perfectly formed volcano on Earth. But I haven't come to see either of these wonders of our planet; I'm travelling to the southeast of the island, to the province of Batangas, to see a very special lake – a place where it is possible to see conditions similar to those that powered the very first life on Earth.

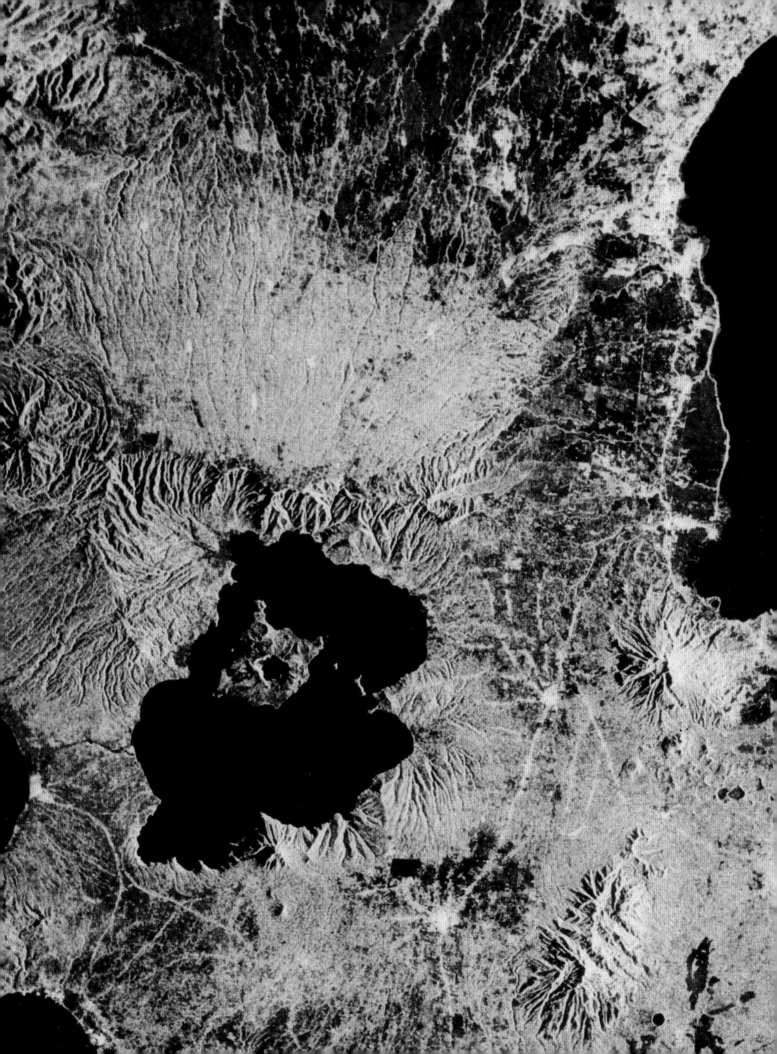

# LIFE'S FIRST
# ENERGY SOURCE

The first day of filming is usually difficult. In parts of the world distant from Britain, there is the time zone to cope with; a seven-hour forward shift in the case of the Philippines. Add the effects of airports and flights, vaccinations and malaria tablets, changes of weather and the general unfamiliar excitement of the distant and exotic, and it's sensible to ease into the four-week trip gently. Imagine our delight, then, when on day one of 'The Fire of Life', we were confronted by the Taal volcano. Its geography is daunting: a lake within a lake; a giant flooded caldera with Volcano Island at its centre – itself home to a second, bubbling flooded crater.

The slopes of Volcano Island are heavily wooded with thick, subtropical vegetation, hacked away along the shores to leave space for fishing villages that play a dangerous and unequal game of poker with a sleeping giant. In the film, you see my arrival on a boat, stabilised in the traditional way with wooden outriggers. Our village, whose little buildings and inquisitive children make an engaging backdrop to the film, might be described as a sort of brightly coloured island paradise, but this would be a dangerous cliché. In 1911, all of the villages on these shores were wiped out and the majority of the population killed by one of the regular eruptions of Taal. The villagers returned, even though settlement along the shores of Taal's inner lake is illegal because of the danger. When we arrived in the autumn of 2011, we required a special permit to visit, but even this would not have been granted the previous summer because volatile gases released by increased seismic activity made the trip too dangerous, even for a gas-masked film crew. The village settlements are, in my view, inexplicably tolerated by the local authorities, but poverty can be a powerful and defiant motivator.

*The crater was excavated in a series of eruptions stretching back around 140,000 years. Over the millennia, 120 billion cubic metres of ash and rock have been blown into the Earth's atmosphere, forming a crater 30 km across and in places 150 m deep.*

**LEFT:** Taal volcano is one of the most dangerous in the world, in terms of threat to life and property. During an eruption in 1965, pictured here, a new explosion crater was formed.

**ABOVE:** Taal Lake has a high sulphuric content, releasing toxic gases. Living on its wooded shores is a risky business.

Taal is one of the world's 16 'Decade Volcanoes', the most dangerous in terms of threat to life and property. It shares this distinction with more infamous names: Etna, Rainier and Vesuvius. Its power is made manifest by the history of the Volcano Island – the isle didn't exist 300 years ago. In the eighteenth century it was part of a bay – a coastal feature that looked out onto the South China Sea. That was before a series of major volcanic eruptions transformed the landscape, closing the circle of land and leaving the Pansipit River as the sole connection to the ocean. The crater lake itself was excavated in a series of eruptions stretching back around 140,000 years. Over the millennia, 120 billion cubic metres of ash and rock have been blown into the Earth's atmosphere, forming a crater 30 km across and in places 150 m deep.

The energy to create this volcanic 'Russian doll' is drawn from an ancient store deep within our planet. The Earth's core is a raging cauldron of molten iron. At 5,500°C, it is hotter than the surface of the Sun. As energy is neither created nor destroyed, this heat must have come from somewhere. The Earth was formed 4.54 billion years ago from a cloud of dust and gas slowly drifting together. Planetary formation begins when rocky objects around 1 km across – known as planetesimals – begin to clump together under their own gravity. These collisions heat up the rocks through the release of gravitational potential energy as they fall together. This heat, released in the Earth's initial gravitational collapse, has remained inside our planet for billions of years. But this accounts for only about half of the 44 terawatts that Earth leaks away into space via the action of plate tectonics, volcanoes and other active geological phenomena. The rest of Earth's heat has an origin more ancient still. It is released by the radioactive decay of elements such as uranium and thorium, which exist in vast quantities in the Earth's upper layers – the lithosphere and mantle. We know of only one place violent enough to create these heaviest of elements – supernova explosions.

Supernovae are the most violent explosions in the Universe – they release so much energy that they can outshine an entire galaxy. The last supernova visible to the naked eye from Earth was the Kepler supernova, which burst into the night sky on 9 October 1604, outshining every other star there, even though it was 20,000 light years away. These rare, spectacular events are both devastating and productive, releasing the vast amounts of energy needed to force lighter nuclei together to make the heaviest elements. Every atom of gold and silver on Earth today, including those locked up in the jewellery you might be wearing while reading this book, was

**BELOW:** Volcanic activity – shown in this image of Kilauea volcano in Hawaii – is one of the mechanisms by which the Earth releases heat into the atmosphere.

*The Earth's heat is released by the radioactive decay of elements such as uranium and thorium. We know of only one place violent enough to create these heaviest of elements – supernova explosions.*

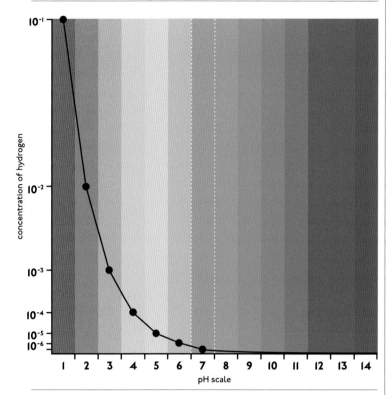

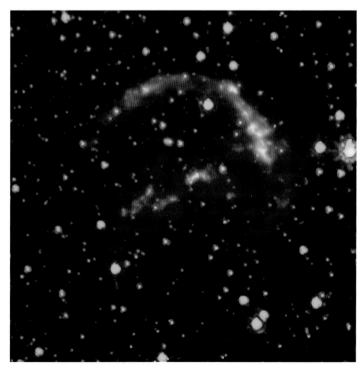

**ABOVE:** The Kepler supernova was first observed on 9 October 1604. Its violent explosion released so much energy that it was visible to the naked eye for several weeks. Its remnant – shown here – is still an object of much study in astronomy.

produced in the last violent moments of a long-dead star. The reason why such extreme conditions are necessary to produce the heaviest chemical elements is that they don't really want to be together. All elements heavier than iron (which is only the 26th out of the 92 naturally occurring elements) would in general be happier in lighter bits. The heaviest elements, such as uranium and thorium, achieve this by naturally decaying into lighter elements given enough time – this is what we call radioactivity. When they do so, they release a small amount of the energy used to form them back into the Universe, because energy is always conserved. It would be correct to think of these heavy nuclei as batteries, storing a tiny fraction of the energy of an ancient supernova explosion.

When the Earth formed, part of the collapsing dust cloud was made up of these radioactive elements, and because they are heavy, they quickly sank towards the planet's core. They have remained deep underground ever since, decaying slowly and releasing ancient stellar power into the beating heart of our planet. Our Earth is powered by nuclear energy, whether we like it or not!

At Taal Lake this ancient energy is continually and visibly being released. It has been a long journey through space and time, but along the inner shores of Volcano Island it reaches the surface, manifesting itself as the boiling water and steam that make this site such a dramatic place. Finally, after many billions of years, this trapped stellar energy will return to space in the form of infrared photons, leaving our planet forever. Not all of that energy escapes just yet, however. Some of it becomes trapped once more in the water of the lake as chemical energy.

Sitting precariously on donkeys, a dripping jet-lagged film crew wobbled inelegantly down the slopes of the inner crater, through the vegetation and biting insects and into the vaporous inner caldera to investigate the composition of the bubbling waters. The simplest of tests reveals the nature of the energy trapped in the lake. Universal indicator paper (litmus paper) is found in every school chemistry lab. For those of you who don't remember (Mr Bell would have clipped you around the ear and grumbled deeply), universal indicator paper measures the pH of a liquid by changing colour. A pH below 7 means the liquid is acidic; above pH 7 means it is alkaline. Tap water is very close to neutral – pH 7. A quick dip of the paper into the central volcano lake reveals that it is pH 3 – a weak acid. The definition of pH is rather complicated, but in essence pH is a measure of the concentration of hydrogen ions (or in other words protons) available for chemical reactions. A lower pH means that there is a higher concentration of free protons floating around in the water. The reason for the acidity of the lake is that the energy released by the volcano melts rocks close to the surface, releasing a host of gasses including sulphur dioxide. When sulphur dioxide bubbles through water it dissolves, forming a weak acid, similar in strength to lemon juice. For the chemistry students, we should also be speaking in terms of the concentration of hydronium ions, $H_3O^+$, but for our purposes the simplification is appropriate. What matters is that a small amount of the volcano's energy has been used to change the concentration of hydrogen ions in the water, and this has the potential to do things – it is a store of chemical energy. More colloquially, it is a battery. ◉

# ON PROFESSOR COX'S BATTERY AND THE ORIGIN OF LIFE

The principle of a battery is a very simple one. It is a device that stores energy in the form of chemical energy, and allows for its release when connected in an electrical circuit. It is not known how long we have been constructing these stalwarts of the modern age, but a collection of intriguing artefacts discovered near Baghdad in the 1930s suggests that electrochemical cells may have been manufactured as far back as AD 225. The precise purpose of the Baghdad Battery remains uncertain, but its design suggests, at least superficially, a remarkably modern function at least 1,500 years before the recognised invention of the battery by Alessandro Volta at the turn of the nineteenth century.

**PROTON WATERFALL**
The production of energy is like a water wheel, sitting in a proton waterfall

ATP

hydrogen

Since then, hundreds of different types of battery have been designed and manufactured, but despite the huge variation in ingredients, size and performance, the principle underlying every battery remains the same. A battery consists of two half cells, one with an abundance of positive ions and the other with an abundance of negative ions. If these two half cells are connected by a bridge that allows ions to flow, but not water, then you have a battery.

In general, physical demonstrations are a difficult thing to do in a modern documentary, because they can leave the presenter – me – looking like James Burke without a safari suit. Mindful of the need for the emperor to remain clothed, we were careful, but the flow of ions is so fundamental to life that the construction of a fuel cell on the inner crater rim of the Taal volcano seemed to me both informative and appropriate.

The simplest fuel cell consists of two bottles connected by a membrane that allows ions to pass, but which is impermeable to liquid.

This demonstration illustrates one of the most important chemical processes in living things: the use of proton gradients, or waterfalls. A high concentration of protons on one side of a membrane can be used to power things – in this case, an electric motor. Proton gradients occur quite naturally on Earth, and Taal's twin lakes are a geologically spectacular example. The acidic inner lake, primed by bubbling volcanic gases, is a reservoir of protons – the top of a waterfall. Taal's outer lake is mildly alkaline, owing to the reaction of the water with

the rocks of the shore. This is the bottom of the waterfall – a lake with a proton deficit. This configuration of lakes is therefore a giant natural battery, storing energy derived from the heat of the inner Earth as a frozen waterfall of protons. If there were a mechanism for unfreezing the waterfall and allowing the protons to flow, this energy could be released and used to do useful work.

Remarkably, virtually all life on Earth breathes using proton gradients. To be more specific, cells manufacture adenosine triphosphate (ATP) – life's battery – using an enzyme called ATP synthase. This biochemical machine is in effect a water wheel, sitting in a proton waterfall and churning out ATP molecules like a mill. Energy released by the oxidation of food is not used directly to manufacture ATP. Instead, it is used to pump protons across a membrane – up a waterfall. At first sight, this seems like a rather convoluted thing to do, but there are good reasons why life goes to all this trouble. The details are, as always in biochemistry, complex, but effectively the use of a waterfall allows a cell to build an ATP molecule in stages, by storing little bits of energy in its proton waterfall until it has enough to carry out the chemical reactions required to synthesise the ATP molecule. The discovery of this process – called chemiosmosis – earned a Nobel Prize for the British biochemist Peter Mitchell in 1978. The important point for us here, though, is not the detail of the chemistry but its ubiquity. Chemiosmosis, along with the use of ATP and the genetic code itself, is common to all life, and this very strongly suggests that it must have been present right back at the beginning. ◉

**FUEL CELL:** A hydrogen fuel cell produces electricity by creating an artificial proton gradient, separated by an electrolyte membrane

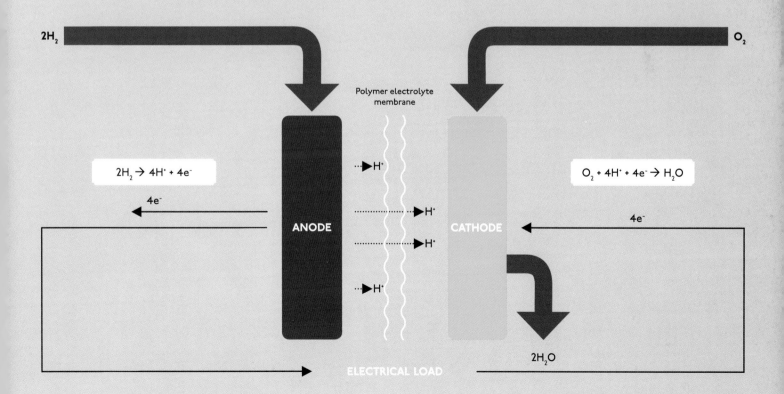

$2H_2$

$O_2$

Polymer electrolyte membrane

$2H_2 \rightarrow 4H^+ + 4e^-$

$O_2 + 4H^+ + 4e^- \rightarrow H_2O$

$4e^-$

$4e^-$

ANODE

CATHODE

$H^+$

$H^+$

$H^+$

$H^+$

$2H_2O$

ELECTRICAL LOAD

*We may never know the precise sequence of events that led to the moment of genesis, but we do know that it happened in an unfamiliar world. For the first few hundred million years after its formation, our planet was a sterile, hot, toxic and turbulent place, scarred by volcanoes and raging seas, and completely at odds with our image of a living planet. And yet somewhere, beneath a newly formed ocean, there was a Garden of Eden. Green and lush it was not, but Eden nonetheless – a place where inanimate inorganic ingredients slipped ineluctably down the chemical slope to complexity.*

**RIGHT:** 'Black smokers', such as this one photographed by DSV *Alvin*, were first discovered in 1977. They are a type of deep-water superheated vent, and they typically emit particles with high levels of sulphur-bearing minerals.

# SEARCHING FOR EDEN:
# A WARM LITTLE POND...

The search for Eden has driven scientists to some of the most remote places on Earth. Geologists and biologists have scoured the planet for evidence of the earliest forms of life, and some of the most extraordinary finds have come from studies in Australia. Marble Bar in Western Australia is a gold-rush town famed for holding the world record for the longest stretch of consecutive days with temperatures above 37°C. But in 2011 it gained a second accolade with the discovery of the oldest verified fossils on Earth. The image (below) of ancient single-celled organisms caught between grains of sand is a glimpse of a simpler world, 3.45 billion years ago. It is not easy to validate a find of this sort – geological irregularities or crystal formations are often mistaken for ancient life. There are discoveries claimed to be older, but this sample – unearthed by Martin Brasier of Oxford University and David Wacey of the University of Western Australia – has so far stood up to scrutiny.

These primordial life forms were found on what was probably an ancient beach, but the conditions were nothing like those on Australia's coastline today. The oceans were acidic, and heated to 45°C – hotter than most of us can stand in the bath. The atmosphere would have been radically different, heavy with volcanic gases and devoid of oxygen. Further analysis by Brasier and Wacey revealed tiny deposits of sulphur crystals in and around the cells that are different to the sulphur compounds naturally occurring in rock, suggesting that these life forms metabolised the plentiful, volcanically released molecules such as hydrogen sulphide, just as some extremophiles do today.

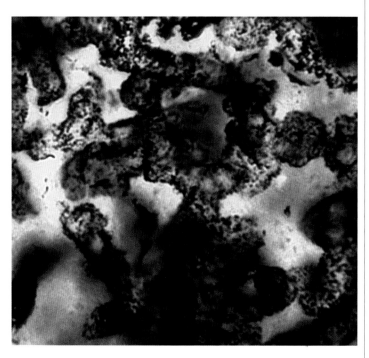

**ABOVE:** At Strelley Pool, ancient single-celled organisms were found in between grains of sand in the 3.4-billion-year-old sandstone.

During the filming of *Wonders*, I've seen some of these life forms for myself, from the hydrogen-sulphide-loving cave-dwelling snottites in the Cueva de Villa Luz in Tabasco, Mexico, to the yellow mats created by organisms living in the super-heated waters of hydrothermal vents 1.6 km below the Sea of Cortez – a whole ecosystem based around sulphur chemistry.

My journey into the Sea of Cortez was aboard a vehicle that has become something of a legend in the history of deep-sea exploration: DSV *Alvin*. *Alvin* is a 17-tonne submersible capable of operating at depths of up to 4.8 km. It was a great privilege for me to journey to the bottom of the sea aboard *Alvin*, separated by just a few centimetres of titanium from 200 atmospheres of water pressure, although it was a routine expedition for *Alvin* and its crew from the Woods

*This image of ancient single-celled organisms caught between grains of sand is a glimpse of a simpler world 3.45 billion years ago.*

Hole Oceanographic Institution, Massachusetts. Since its launch in 1964, *Alvin* has explored some of the most extreme environments on Earth, including the wreck of the *Titanic*. But in the search for the origins of life, perhaps none of its dives has been more important than an expedition mounted in the mid-Atlantic in 2003.

Although some early life forms would have used sulphur chemistry to power their metabolic processes, they were almost certainly not the first living things. There is as yet no fossil evidence or visible trace of life before the Marble Bar fossils, but that doesn't mean we can't infer what the earliest life might have been like. We begin with a possible cradle, an echo of Eden, discovered during an extraordinary series of expeditions with Alvin.

In December 2000, deep below the surface of the Atlantic Ocean between Bermuda and the Canary Islands, *Alvin* embarked on a voyage of discovery of an extraordinary underwater landscape. The primary purpose of this mission was to explore the Atlantis Massif, a prominent 8,200 m mountain range stretching over 16 km along the ocean floor, but the most important scientific find was a new type of deep-sea vent, unlike anything seen before. Three years later, the *Alvin* team returned to explore these vents and bring samples back to the surface. They discovered a unique biochemistry, and a unique world, which they named the Lost City.

Resting on the Atlantis Massif are 30 towers of calcium carbonate, some 60 m high, formed by hot water, minerals and gasses rising up from the deep Earth. These vents are quite unlike the sulphurous black smokers I saw in the Sea of

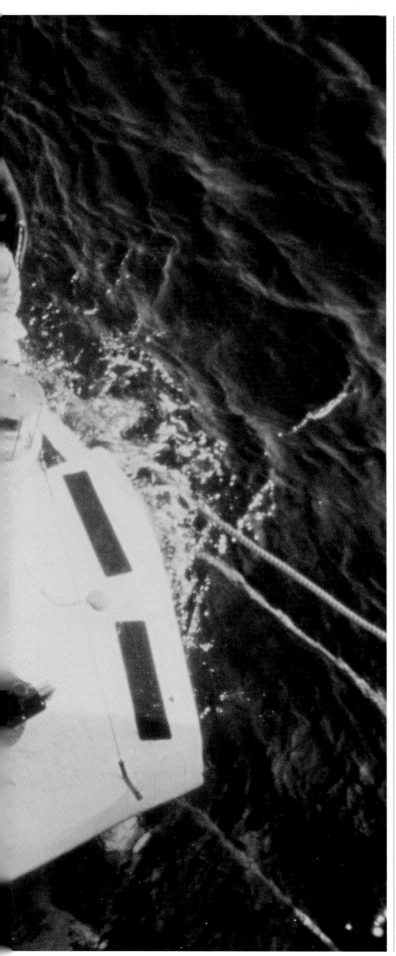

**LEFT:** DSV *Alvin*, a manned deep-ocean research submersible, prepares to dive. The purpose of its mission was to explore the Atlantis Massif, a prominent mountain range deep below the surface of the Atlantic Ocean.

Cortez. Those vents are superheated, with water temperatures rising beyond 300°C. In the Lost City, the waters are a relatively balmy 90°C. But the critical difference for the study of origins of life is the chemical reactions that occur between the heated water and the peridotite rock lining the sea floor. These reactions produce methane and hydrogen gas, and not the carbon dioxide and hydrogen sulphide characteristic of the black smokers. So while the black smokers produce an acidic environment, here in the Lost City, the opposite happens: the seawater becomes highly alkaline (pH 9–11).

This is important because we know that when life began, Earth's oceans were mildly acidic. This means that there would have been naturally occurring proton gradients throughout the rocky chambers of the vents – proton-deficient water surrounded by a proton-rich ocean. The vents would also have been rich in organic materials and minerals such as iron and nickel, the raw materials of life, which are held in high concentrations in porous chambers suspended in the middle of a powerful, naturally occurring proton waterfall. Even today, these oxygen-free chambers are still lined with archaea known as Lost City Methanosarcinales, which use methane as their metabolic fuel.

These vents provide the most compelling scientific vision of Eden: high concentrations of organic materials, held in what chemists would call the far-from-equilibrium conditions of proton waterfalls, mean that complex chemistry emerges quite naturally. And, so the theory goes, this is the reason why all life on Earth today shares the same predisposition for proton gradients. It always did! Our common ancestor was not a cell, nor even some kind of simpler free-living thing, but a set of chemical reactions occurring inside a small chamber of rock, rich in organics and lined with naturally occurring catalysts, suspended in a naturally occurring proton waterfall powered by the inner heat of the Earth. And when the time came for life to leave Eden, it did the simplest thing it could – it put a bag around the already established chemistry and floated away. And quite wonderfully, the echoes of our origins are still present today in every living cell in our bodies, and in every animal, plant, alga, insect, bacterium and archaeon. We all carry the chemistry of the primordial Earth around with us, carefully and meticulously re-created by the same biochemical machinery that requires it to survive.

It sounds like a story too outrageous to be true, but the evidence lies within us all. ◉

# UNIVERSAL LIFE

**BELOW:** A micrograph of the bacteria staphylococcus. Every living thing consists either of a single cell, or colonies of cells working together.

**RIGHT:** This coloured transmission micrograph shows a longitudinal section through a healthy heart muscle. A high concentration of large, oval mitochondria is visible between the muscle fibres (pink).

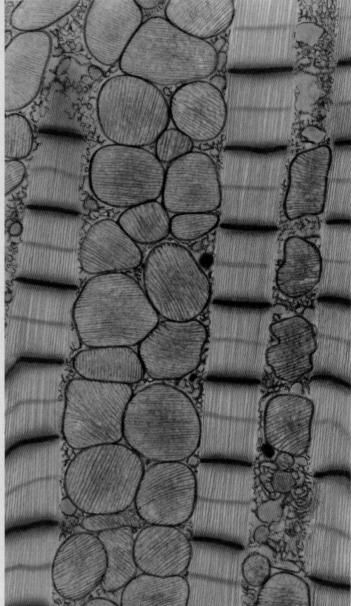

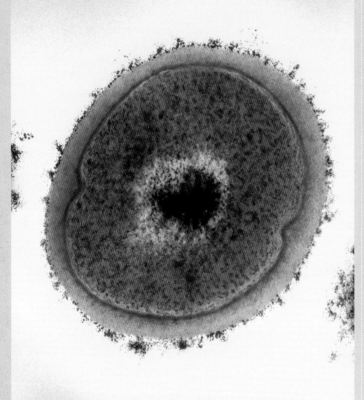

Each one of us contains about 50 trillion cells, working together to create the complex structures of the human body.

Every living thing consists of either a single cell, or colonies of cells working together, from the simplest organisms to the multicellular intricacy of every plant and animal on Earth. It took the invention of the microscope to reveal what was invisible to the naked eye: the 'atom' of life, the smallest living biological structure, discovered and named by Robert Hooke in 1665. For the last 150 years we have been exploring the micro-universe contained within these cells, uncovering the common chemical processes and landscapes of complexity that underpin the workings of all life.

Cells are divided into two distinct types: eukaryotes and prokaryotes. Prokaryotic cells are the simpler of the two, containing far less cellular machinery and, crucially, lacking a nucleus. This is the structure of the simplest life forms on Earth – the bacteria and archaea. Prokaryotic organisms are almost always unicellular, and provide

the clearest living example of how life must have looked and lived during the first 2 billion years of life on Earth. Eukaryotic cells, in comparison, are far more complex machines that emerged only around 2 billion years ago. These cells are the building blocks of human beings and every creature on Earth that we would regard as complex – from fungi to plants, from protists to animals. Compared to the simplicity of bacteria or archaea, the eukaryotic cell appears like a twenty-first-century factory, full of advanced structural technology, compared to an artisan's workshop. As well as a nucleus containing all the cell's chromosomal DNA, the inside of a eukaryotic cell is packed full of other functional units. The most striking for our story – a crystalised echo of the eukaryotes' evolutionary past – are the mitochondria.

Mitochondria are tiny organelles, around a micron in diameter, found in virtually all eukaryotic cells. Those very few cells that don't possess them probably did at some point in their past. Some cells contain just one mitochondrion,

**BOTTOM:** A transmission micrograph of a section through a mammalian cell, showing the nucleus (pink), the nucleolus (dark brown) and the cell cytoplasm (green). The small brown bodies in the top of the cell are mitochondria.

**OVERLEAF:** A transmission micrograph of a sectioned mitochondrion from a fat cell. In some fat cells, specialised mitochondria help to generate body heat.

**BELOW:** A transmission micrograph of onion root cells, showing the nucleus, leucoplasts, mitochondria and cell walls. Mitochondria are found in virtually all eukaryotic cells.

**RIGHT:** This transmission micrograph shows a mitochondrion in a plant cell. This is where the proton waterfalls spin the turbines of ATP synthase to produce ATP, the 'universal battery of life'.

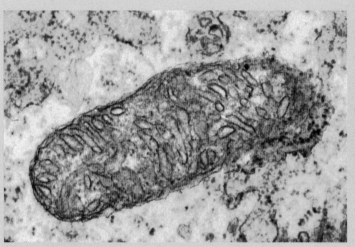

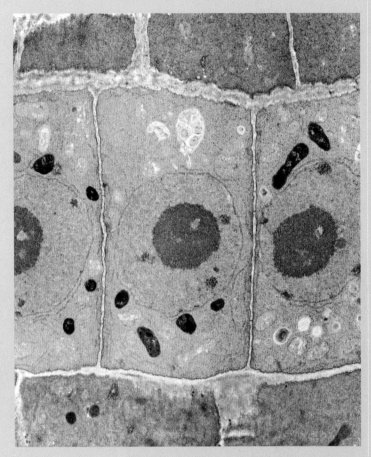

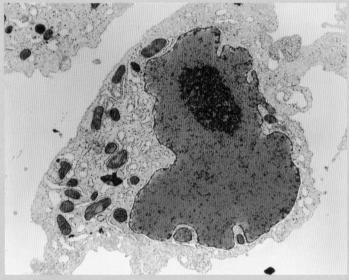

but many contain hundreds of these discrete units. Inside each of them is a maze of compartments separated by a series of membranes. Within these spaces the mitochondria perform their crucial function, which is to power the cell. This is where the proton waterfalls spin the turbines of ATP synthase to produce ATP, the universal battery of life. The mitochondria are the seat of the production of ATP by the oxidation of food in every living thing. This is where over 80 per cent of your energy comes from. Although you can produce ATP anaerobically, for example through a process known as glycolysis in your muscles, this won't keep you going for long. Without mitochondria, there would be no humans. In fact, there would in all probability be no complex life on Earth at all. That's a strong statement, but to see why it's probably a valid one, we need to look a little more carefully at the nature of the mitochondria themselves.

The most striking property of the mitochondria is that they have their own DNA, stored as a loop. This looks exactly like bacterial DNA, because it is. The mitochondria are bacterial in origin – they exist symbiont-like inside our cells. What led these ancient bacteria to end up inside another cell in the first place? This is one of the great open questions in biology. Endosymbiosis, as the process is known, has certainly happened more than once in the history of life on Earth. We've already seen an example in Chapter 1: the chloroplasts inside all green plants and algae were once free-living cyanobacteria. But the fusion of an ancient cell with a bacterium skilled in control of proton gradients and the efficient production of ATP via aerobic respiration may have been even more important because it has been proposed that this was the origin of the eukaryotic cell itself. Recent DNA studies of modern cells suggest that the original host cell for the first mitochondria may have also been a prokaryote – an archaeon. Further, it is proposed that the subsequent development of the internal complexity that characterises eukaryotic cells, and the organisms composed of them, may have been possible only once the immense energetic

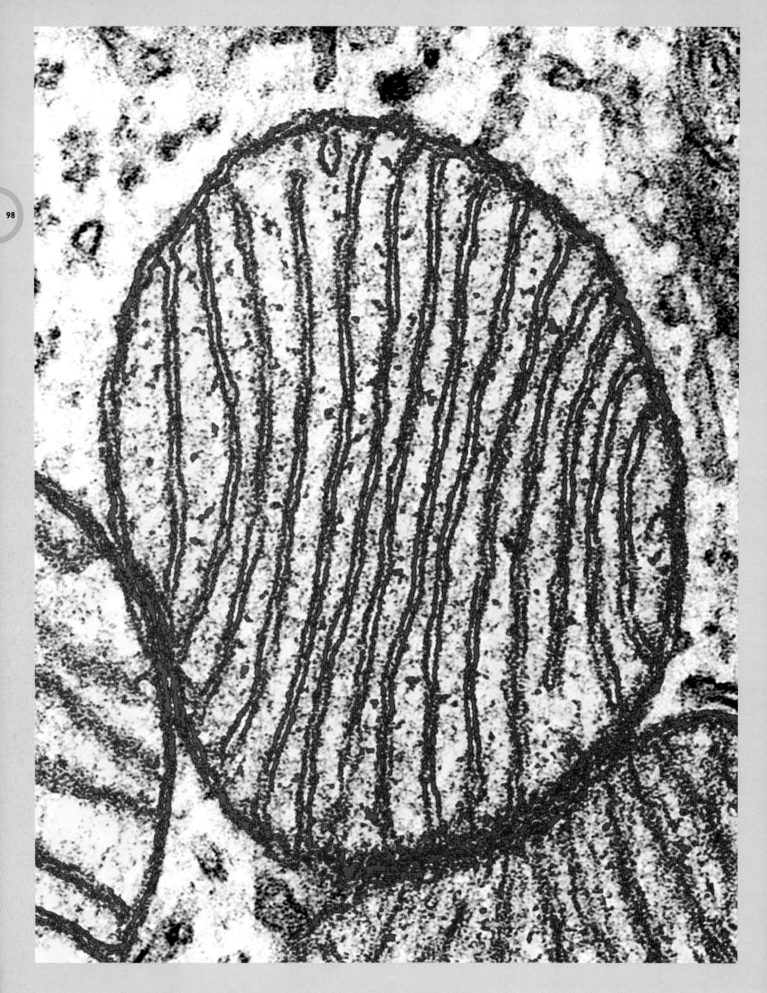

advantage provided by the hosting of mitochondria became available. If this is true, then the reason life on Earth was so 'dull' for 2 billion years (with apologies to the bacteria and the archaea, which, lets face it, would never have made great dinner guests) is that a cell with a mitochondrial powerhouse is a prerequisite for the complexity of the eukaryote.

Drawing all this together, we have a wonderful story. We should say that this is at the cutting edge of research, and some of it may be wrong, but it is a possible description of the origin of life on Earth, supported by evidence. Life began as a series of chemical processes in the rich organic environment of deep ocean alkaline vents. Naturally occurring proton gradients provided the power source, and these waterfalls were intimately connected with the biochemistry of life from the start. When life came to leave the vents, it took that chemistry with it. For 2 billion years, life flourished but remained simple because of energetic constraints. Prokaryotes do not have the machinery needed to produce a bounty of ATP, a necessary luxury for the evolution of complex, multicellular life. Then, in an event that may have had an almost vanishingly small chance of success, an archaeon 'swallowed' a bacterium, and both survived. In its protected new environment the archaeon became highly specialised, paring its functions down to focus solely on the efficient production of ATP from a proton waterfall. Given another 2 billion years, the descendants of that chimera have succeeded in working all that out, not least because the evidence is written into the workings of every cell in their bodies. The mitochondria, with their chemistry echoing early Earth, and their rings of DNA betraying their bacterial origins, are the signpost pointing back to the origin of life in deep ocean vents, powered by energy locked away from Earth's formation and the deaths of ancient stars. What a beautiful thought. ◉

## ATP PRODUCTION: THE UNIVERSAL BATTERY OF LIFE

The production of ATP, which cells use to store energy, is a very complex process that takes place in the mitochondrial membrane. Anachronistically ATP production requires the consumption of energy, in the form of two ATP molecules, to produce energy, in the form of four ATP molecules.

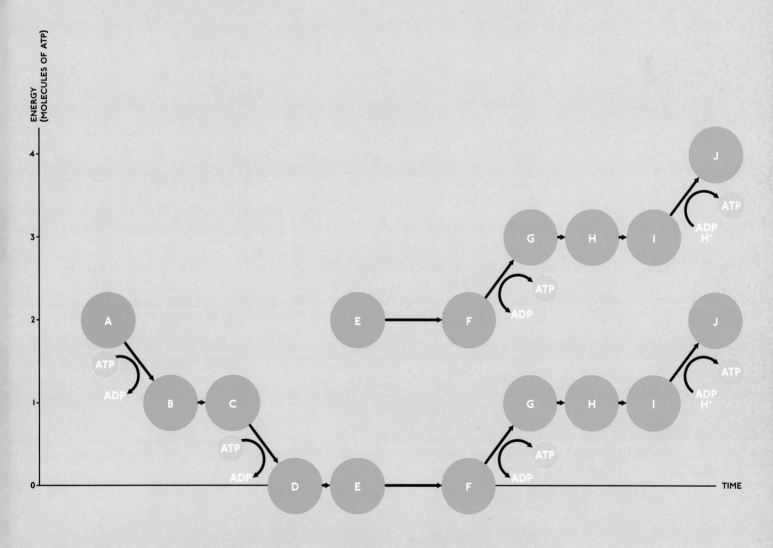

| A | Glucose | C | Fructose 6-phosphate | E | Glyceraldehyde 3-phosphate | G | 3-phosphoglycerate | I | Phosphoenolpyruvate |
| B | Glucose 6-phosphate | D | Fructose 1,6-bisphosphate | F | 1,3-bisphosphoglycerate | H | 2-phosphoglycerate | J | Pyruvate |

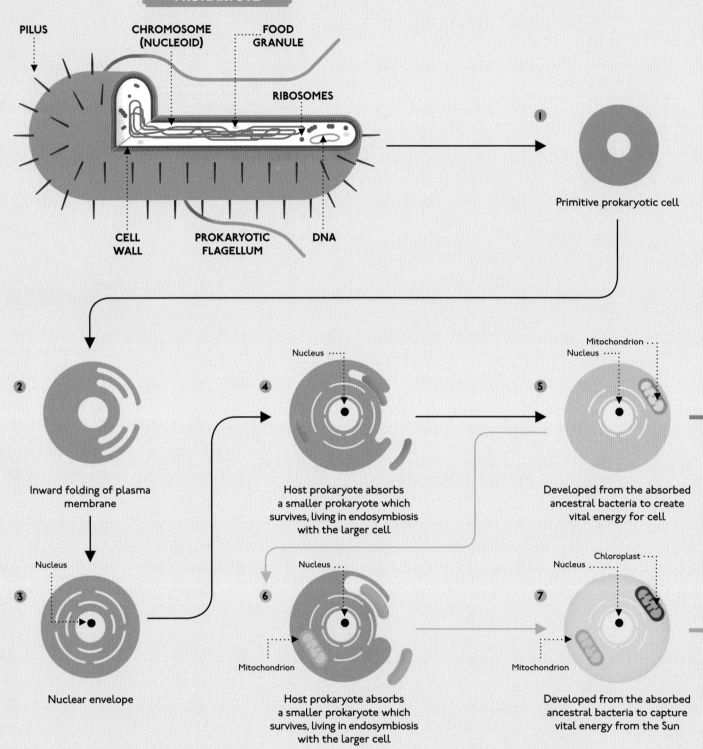

PROKARYOTE

PILUS

CHROMOSOME (NUCLEOID)

FOOD GRANULE

RIBOSOMES

CELL WALL

PROKARYOTIC FLAGELLUM

DNA

① Primitive prokaryotic cell

② Inward folding of plasma membrane

③ Nucleus
Nuclear envelope

④ Nucleus
Host prokaryote absorbs a smaller prokaryote which survives, living in endosymbiosis with the larger cell

⑤ Mitochondrion
Nucleus
Developed from the absorbed ancestral bacteria to create vital energy for cell

⑥ Nucleus
Mitochondrion
Host prokaryote absorbs a smaller prokaryote which survives, living in endosymbiosis with the larger cell

⑦ Chloroplast
Nucleus
Mitochondrion
Developed from the absorbed ancestral bacteria to capture vital energy from the Sun

# PROKARYOTIC AND EUKARYOTIC LIFE

Cells are divided into two main classes. Prokaryotic cells (primarily bacteria) lack a nucleus, whereas eukaryotic cells have a nucleus in which the genetic material is separated from the cytoplasm. Prokaryotic cells are generally smaller and simpler than eukaryotic cells and first emerged about 3.8 billion years ago, whereas eukaryotic cells evolved much later (only about 1–1.5 billion years ago). In addition to the absence of a nucleus, prokaryotes do not contain cytoplasmic organelles or a cytoskeleton, and their genomes are less complex. In spite of these differences, all life uses the same basic molecular mechanisms, showing that all cells are descended from a single primordial ancestor.

# ANIMAL EUKARYOTE

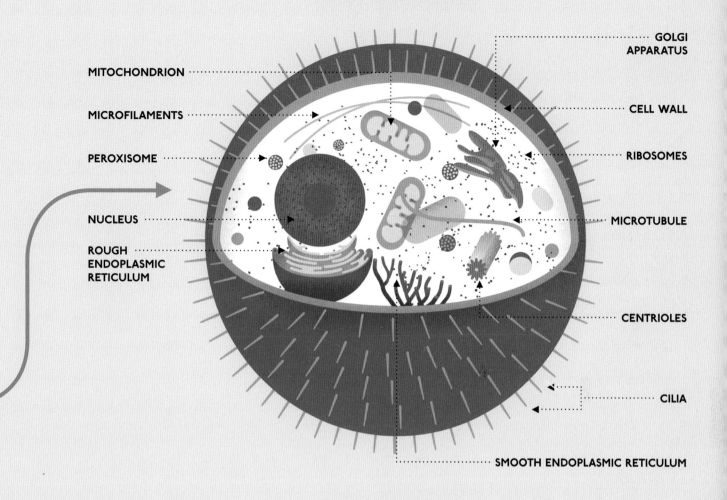

MITOCHONDRION

MICROFILAMENTS

PEROXISOME

NUCLEUS

ROUGH
ENDOPLASMIC
RETICULUM

GOLGI
APPARATUS

CELL WALL

RIBOSOMES

MICROTUBULE

CENTRIOLES

CILIA

SMOOTH ENDOPLASMIC RETICULUM

# PLANT EUKARYOTE

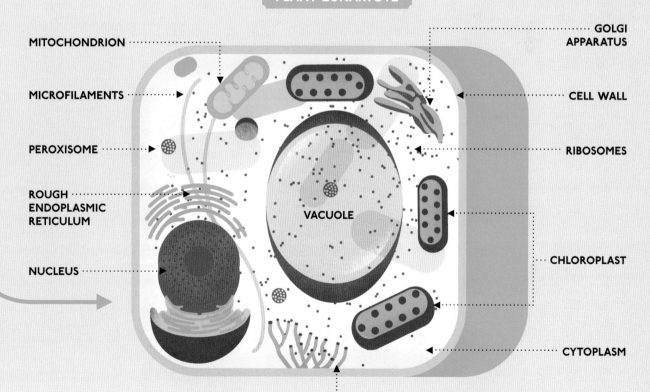

MITOCHONDRION

MICROFILAMENTS

PEROXISOME

ROUGH
ENDOPLASMIC
RETICULUM

NUCLEUS

VACUOLE

GOLGI
APPARATUS

CELL WALL

RIBOSOMES

CHLOROPLAST

CYTOPLASM

SMOOTH ENDOPLASMIC RETICULUM

# LIFE AND THE SECOND LAW OF THERMODYNAMICS: SCHRÖDINGER'S PARADOX

We have looked in some detail at the energetics of life, and even glimpsed a candidate for its volcanic spark, deep beneath the oceans of the young Earth. But a power source alone does not make a living thing. In *What is Life?* Schrödinger notes that a defining feature of a living thing is that it avoids decay, which is the reason why organisms appear 'so enigmatic'. To a physicist, this ability is perhaps even more surprising, because physicists have the second law of thermodynamics engraved on their consciousness at an early stage of their education. Recall that the first law deals with the conservation of energy; energy can neither be created nor destroyed. The second law deals, in a sense, with what can be done with that energy, and it introduces a notoriously tricky physical quantity called entropy.

Entropy has a simple definition: at the absolute zero of temperature, which is just under –273°C, all substances have zero entropy. If you add heat energy to a substance in small steps, the change in its entropy is equal to the sum of all the small amounts of heat you add in each little step, divided by the temperature at which each little bit of heat is added. This doesn't sound like a particularly useful or exciting quantity, although at least it is well defined, yet it is absolutely fundamental to the thermodynamics of anything, including life. To see why, there is an alternative and equivalent definition of entropy by the nineteenth century Austrian physicist Ludwig Boltzmann. His equation is written on his grave in Vienna:

$$S = k \cdot \log W$$

In words, this equation says that the entropy ($S$) is proportional to the number of ways the component parts of something can be arranged ($W$) such that it is not changed; $k$ is a constant known as Boltzmann's Constant. For completeness, it is equal to $1.3806505 \times 10^{-23}$ Joules per Kelvin. This sounds a bit woolly, but it's equivalent to the definition in terms of heat and temperature. The important thing is that this definition of entropy is concerned with how well ordered something is.

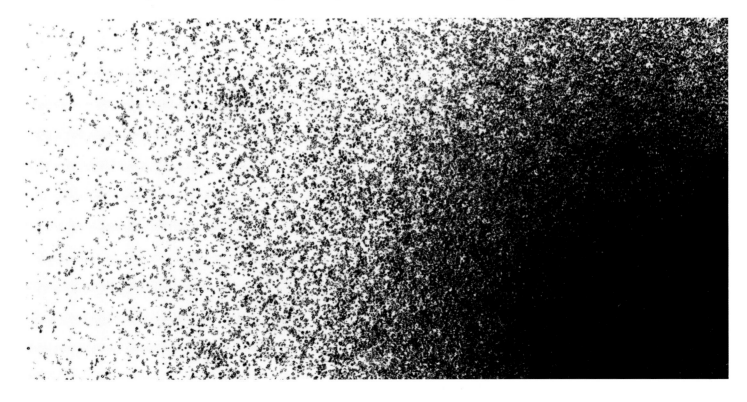

Think of a teacup. It is made up of lots of atoms, arranged into the shape of a teacup. Now imagine smashing the teacup so completely that every one of its atoms becomes disconnected from every other, and piling up the atoms in a little heap. Entropy is concerned with the number of ways you can arrange these atoms to make the teacup and the heap. A teacup is a very specific arrangement of atoms; if you change the position of too many of them, it will no longer be a teacup. A heap, in contrast, is a rather less specific arrangement. You could swirl the heap of atoms around and be left with another heap that is for all practical purposes unrecognisable from the first one. Because there are more ways of arranging atoms into a heap than a teacup, Boltzmann's equation tells us that the entropy of a heap is larger than the entropy of a teacup (because $W$ is bigger).

The second law of thermodynamics states that the entropy of an isolated system always stays the same or increases – it NEVER decreases. This is obvious when rephrased in the language of teacups. A pile of atoms will never assemble themselves into a teacup if left alone or even gently agitated for a billion years, and a tremendous amount of work would have to be done to build such a well-ordered thing as a teacup. In contrast, it is relatively easy to turn a teacup into, if not a pile of atoms, then a highly disordered mess on the floor. This is a universal law of physics – things tend to get more disordered, because it is overwhelmingly more likely for them to do so.

Life is a notable exception. At first sight, it appears to be second-law defying; it 'avoids decay', as Schrödinger wrote, which is perhaps an understatement. Not only does it avoid decay, it actively assembles itself into extremely complex structures. No wonder it is tempting to attribute the properties of living things to a designer or supernatural action; it seems there must be a watchmaker to build a watch, although, as Richard Dawkins put it, the watchmaker will turn out to be blind and most definitely operating in accord with the laws of physics.

This ability to build complexity spontaneously – to lower the entropy of a particular group of atoms – might be taken as the defining property of life; it has been able to create structures as ordered as human beings from the chaos of a primordial, volcanically heated ocean. So it has, but let us emphasise again that it has not broken any physical laws in the process. If it had, then Schrödinger's programme of accounting for life using the laws of physics and chemistry would be defeated. How, then, does life assemble itself in seeming contradiction of the second law? This has become known as Schrödinger's paradox. ◉

# FOLLOW THE SUN

Eight hundred kilometres east of the Philippines, the Island Republic of Palau is one of the smallest sovereign states on the planet, and one of the most beautiful. Consisting of eight principal islands and more than 250 smaller ones, it contains an extraordinary set of landscapes, from the mountainous main island of Babeldaob to the hundreds of low-lying islands and spectacular coral reefs spread out into the Pacific. It is truly an island paradise.

We came to film an animal unique to a single lake on Eil Malk – one of Palau's tiny outlying islands. Shaped like a letter Y, Eil Malk's 19 sq km are uninhabited and densely wooded. It was formed from the remains of ancient coral reefs that were violently pushed to the surface, creating a beguiling, chaotic fragment of land pockmarked with small lakes. At its eastern tip lies a saltwater lake, connected to the sea and yet isolated from it, as the rock channels that empty to the ocean are too small for all but the tiniest marine creatures to navigate. The lake is therefore effectively isolated; an ecosystem sealed for 12,000 years – plenty long enough for its inhabitants to have begun diverging from their ocean-going cousins swimming just a few hundred metres to the east.

We arrive before dawn and, as the Sun climbs above the eastern ridge, head out in canoes to the emerging illuminated strip of water along the western bank. There are just a few individuals at first – shifting golden shapes beneath the greening surface. But as eyes become tuned to the light, the lake takes on an organic, textured density. The water itself seems alive. With apprehension tempered artificially by a wetsuit, I roll backwards off the boat, and land in jellyfish soup.

Eil Malk is home to 23 million golden jellyfish, a subspecies of the spotted jellyfish found in the surrounding lagoons. Isolated from predators, they have a greatly reduced sting, although I notice that they still have the ability to numb the lips as they brush past. They have also lost their spots, and acquired their distinctive golden domes. Most jellyfish, including the spotted jellyfish, obtain their energy from a diet of tiny marine creatures known as zooplankton, which they manoeuvre into their mouth using specialised tentacles known as oral arms. The golden jellyfish, however, have much shorter oral arms. It is difficult at first sight to see why such an adaptation would benefit the animal, but if you observe the behaviour of the golden jellyfish for a few hours, the reason becomes clear – they are far less reliant on zooplankton for food.

The reason the golden jellyfish congregate in shifting illuminated strips on the water is that they are following the Sun, because they rely directly on photosynthesis for nourishment. In a sense, of course, we all do. At the base of virtually every food chain on Earth today is a photosynthetic organism – a tree, a plant, an algal cell or a cyanobacterium.

*Golden jellyfish congregate in shifting illuminated strips on the water because they are following the Sun, relying directly on photosynthesis for nourishment.*

**TOP LEFT:** The golden jellyfish of the saltwater lake on Eil Malk, Palau, follow the path of the Sun, relying mostly on photosynthesis, rather than on zooplankton, for nourishment.

**ABOVE:** Golden jellyfish host photosynthetic algae as symbionts inside their translucent domes. They 'dance' around the lake in order to keep the algae illuminated.

Small animals eat the photosynthesisers, and larger animals eat the photosynthesisers and the smaller animals. What sets the golden jellyfish apart is that they cut out the middleman by hosting photosynthetic algae as symbionts inside their translucent domes. Their complex daily dance around the lake is driven by the need to keep their algae illuminated, while staying away from the shoreline where their only predators – endemic white 'Medusa-eating' sea anemones – await. This results in congregations of jellies migrating around the lake with the shifting shadowline. Diving with them was truly one of the most wonderful experiences I had in the making of the series. It is natural, at least for me, to feel uncomfortable floating in a dense jellyfish soup, but the apprehension is quickly overcome by the deliberate, delicate rotation of these golden domes as they democratically expose their algal symbionts to the Sun.

The algae are from the genus Symbiodinium – common, single-celled algae also found living as symbionts in corals and anemones. The jellyfish engulf the algae as juveniles and by adulthood they make up around 10 per cent of their biomass. Grouped into clusters, they live inside the fibrous cells that crisscross the mesoglea – the jelly-like substance that makes up most of the body of a jellyfish. This symbiotic relationship shows us how, at a fundamental level, life builds order in accord with the laws of nature. ◉

**BELOW:** Jellyfish are free-swimming marine animals consisting of a gelatinous umbrella-shaped bell and trailing tentacles. The bell can pulsate for locomotion, while stinging tentacles can be used to capture prey.

**BOTTOM LEFT:** This deep-water jellyfish (*Ptychogena*), here photographed in Monterey Bay, California, can be found at depths of 50 to 1,200 m. The four broad, bright gonads, held in a cross pattern on the radial canals, are its most obvious characteristic.

**RIGHT:** This luminescent jellyfish (*Pelagia noctiluca*) can be found in the Mediterranean. In Latin, Pelagia means 'of the sea', nocti means 'night' and luca means 'light'.

**BOTTOM RIGHT:** *Chrysaora quinquecirrha*, the Atlantic sea nettle, is frequently seen along the east coast of the USA. Its tentacles inject toxins capable of killing smaller prey or stunning perceived predators.

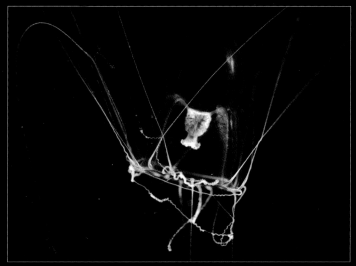

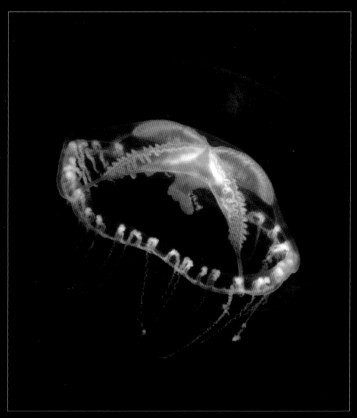

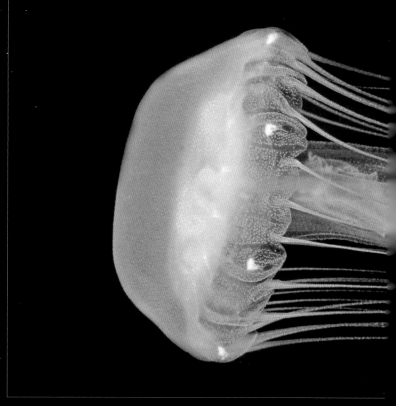

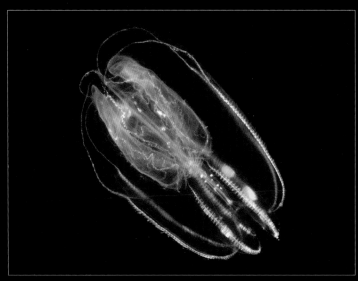

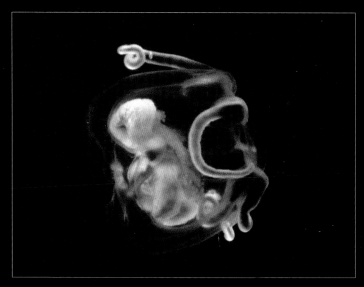

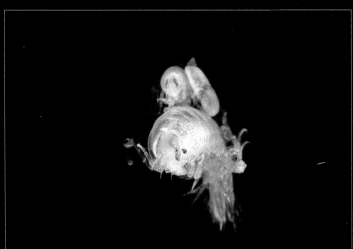

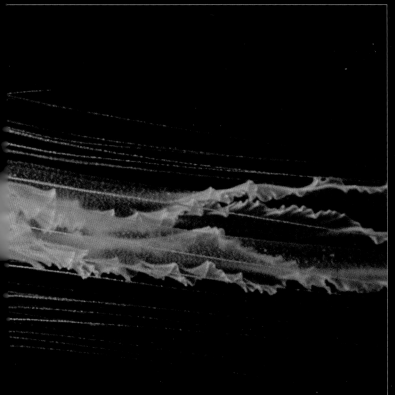

**TOP LEFT:** This comb jellyfish (*Bolinopsis infundibulum*) takes its name from the combs of hair-like cilia, which it beats to propel itself through the water.

**ABOVE LEFT:** *Sarsia tubulosa*, the clapper hydromedusa, with small crustaceans (*Hyperia galba*) visible inside it.

**TOP RIGHT:** A jellyfish medusa found off Heron Island in the southern Great Barrier Reef, Australia.

**ABOVE:** Captured in trawler nets in the Gulf of Mexico, this mesopelagic jellyfish can be found at depths of 450 to 600 m.

# THE ORIGIN OF LIFE'S ORDER

W e studied oxygenic photosynthesis in some detail in Chapter 1. For our purposes here, the biochemical details are irrelevant; we are interested in the thermodynamics. Looking at the iconic equation once more, we see that photosynthesis is a process that builds a complex molecule out of simpler ones – glucose from carbon dioxide and water.

$$6CO_2 + 12H_2O \longrightarrow C_6H_{12}O_6 + 6O_2 + 6H_2O$$

*Energy from the Sun*

Glucose is a relatively intricate structure, certainly in comparison to carbon dioxide and water, and as this molecule is the base of virtually every food chain on Earth, if we can understand how life assembled this complicated thing from simpler components, we will be on our way to understanding the resolution of Schrödinger's paradox.

The answer is in fact quite straightforward. The second law of thermodynamics, which states that entropy (disorder) must always either stay constant or increase in any natural process, applies strictly to the whole Universe, and not just a part of the Universe, unless that part is totally isolated from everything else – which is in practice (and even in principle in

**GLUCOSE**

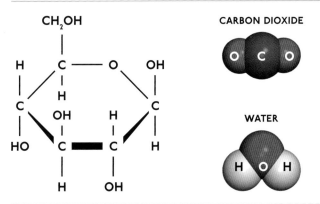

quantum theory) impossible. To be specific, when we consider the thermodynamics of photosynthesis, we must consider both the incoming energy from the Sun and the outgoing energy released as heat during the chemical reaction, as well as the carbon dioxide, water and glucose. If we do that, we will find that the entropy of the whole system, including all the incoming photons from the Sun and all the outgoing photons produced as heat, will indeed rise. But the entropy of a little subsection of the whole thing – the atoms that make up the glucose – will fall. This is the key point. The Sun is a source of ordered energy that is converted by photosynthesis into more disordered energy. In the process, a little of that order is 'borrowed', and used to make glucose.

There is no mystery here; seen as a whole, the entropy of the Universe is increasing, but a little machine such as a chloroplast can do work, using energy supplied to it in a useful form from the Sun, to build an ordered thing, as long as it increases the disorder around it. This is part of the answer to Schrödinger's paradox.

There is, though, a deeper question lurking here. The thermodynamics may all work out in the sense that energy is conserved and entropy always increases, but surely machinery is required to 'extract' the available order in the sunlight and store it as glucose? This is certainly true in the case of oxygenic photosynthesis – the construction of glucose from carbon dioxide and water is a complex problem to solve.

ocean. These gradients can provide the thermodynamic imperative for simple chemicals to assemble themselves into more complex ones. In the gradient-rich and ingredient-rich environments of vents, we know that complex molecules such as acetyl thioesters and pyruvate are formed. These are molecules, more complex than glucose, that are used at the heart of life's metabolic processes today. The emergence of complexity, therefore, is not a mystery. It is a feature of systems that are 'far from equilibrium', as the jargon goes – places where there are waterfalls to power the building process. There is even a re-statement of the second law of thermodynamics, due to Eric Schneider and James Kay, which suggests that systems will utilise 'all available means to resist externally applied gradients'. The assertion is that the emergence of complex systems speeds up the equalisation of temperature gradients, pH gradients, or indeed any gradients, thereby helping the second law on its way to maximising the entropy of the Universe. Gradients don't last long in nature as they soon balance out; the ultimate complex system – life – may be just another consequence of this fundamental truth.

The final part of this paragraph is speculative, sitting in the new and fast-moving world of a subject known as non-equilibrium thermodynamics. Taken at face value, it suggests that the emergence of life is an inevitable consequence of the laws of classical thermodynamics; complexity will emerge in the presence of gradients, and there is a thermodynamic imperative for it to become more complex, irrespective of the details of anything else, including evolution by natural selection. Whether or not this strong statement is correct, it is difficult to say. I think it probably is. But a weaker statement is certainly true: the emergence of complex life is in accord with the laws of thermodynamics, and there is no paradox hiding in the emergence of our magnificent complexity.

This is not the end of our story, however. In *What is Life?* Schrödinger noted that there are two questions arising from life's intriguing order. We have seen how ordered structures emerge quite naturally, in accord with the laws of thermodynamics, in non-equilibrium conditions. But there is a second question that is equally pressing if we are to seek a full explanation for the complexity of the living things we see today. How did the first, relatively simple biological molecules gradually ratchet up their complexity? How did simple living things come to assemble that most spectacular monument to emergent complexity, our 4-billion-year-old ecosystem? Clearly order has been built on order, year by year, millennium by millennium. But how? Schrödinger answered that there is memory in the system – once a complex process has emerged, it need not be continually reinvented. He didn't know what the memory mechanism was, but he deduced a great deal of its properties. He called it an 'aperiodic crystal', a molecule that must be unusually stable and be able to pass information from one generation to the next. Inspired partly by Schrödinger's little book, Watson, Crick, Wilkins and Franklin discovered the precise structure of that aperiodic crystal a decade later in 1953 – the double helix of DNA. And it is to DNA that we now turn. How is it that a molecule is able to encode precious order from generations past and pass it on faithfully to the next? ◉

So if we are to solve Schrödinger's paradox, we must explain how the whole thing could have started without complex machinery – how these thermodynamic machines emerged spontaneously in some deep-sea vent 4 billion years ago. Are there inanimate chemical reactions that build more ordered things out of simple building blocks, which could be used to kick-start the whole process? The answer is yes, in the presence of gradients (or waterfalls, as we've been calling them). And now, everything should start to come together.

Life began in an environment that was out of equilibrium; there were naturally occurring gradients. In the case of the ocean vents, these were proton and temperature waterfalls – hot, alkaline water in contact with a cold, acid

# ONE BIG FAMILY

The Malaysian state of Sabah, on the northern tip of the island of Borneo, is one of the most biodiverse locations on the planet. Home to 15,000 plant species, around 3,000 species of trees, 420 species of bird and 222 species of mammals, it is a place to marvel at the sheer diversity of life on Earth.

Yet for all of its diversity the lives of each individual organism that live in these fertile rainforests are all over in the briefest of time. The life of an insect such as the beautiful dragonflies that flourish all over these lands can be measured in days, mammals such as the great Asian elephants rarely survive for more than a handful of decades, and even the longest-lived organisms – the majestic trees that shape this habitat – are thought to live for no more than a thousand years. Each of these life forms plays nothing more than a cameo in the great story that has unfolded in these forests across tens of millions of years. Every living thing on our planet must die, slavishly adhering to the first and second laws of thermodynamics that govern and guide our Universe. But despite this fleetingness, each species does live on. It is the ability of DNA to conserve information across the generations that has allowed life to develop despite the inevitable progression of decay. It is this preservation of information that has enabled complexity to emerge and be sustained on

**BELOW:** Even where life is abundant, such as in the rainforests of Borneo, all life must adhere to the first and second laws of thermodynamics.

**RIGHT:** Even the largest animals, such as this Indian elephant (*Elephas maximus indicus*), live for no more than a handful of decades. However, their DNA is passed on through subsequent generations.

*Mammals like the great Asian elephants rarely survive for more than a handful of decades, and even the longest-lived organisms – the majestic trees that shape this habitat – are thought to live for no more than a thousand years.*

our planet from the humblest of beginnings 4 billion years ago to the breadth of diversity we see today. DNA is the thread through which every life is connected, not just in the distant past but in the present as well.

Some of our closest genetic relatives can be found living in the rainforests of the Sepilok Forest Reserve in Borneo. Orang-utans are highly adapted for life in the forest canopy, and it is this arboreal lifestyle that is reflected in the distinctiveness of their anatomy. Their arms are twice as long as their legs, and all four limbs are incredibly flexible, each one ending in a hand whose curved bones are perfectly adapted for gripping branches. We share with these beautiful creatures an ancestor that lived somewhere between 15 and 20 million years ago – an ancestor of all the great apes (including ourselves) that we see living on our planet today.

Superficially we might look markedly different to these distant cousins, but look inside each of their cells and we see a very different story. Sequenced in 2011, the orang-utan became the third species of hominid after humans and chimpanzees to have its genome completely sequenced. A captive orang-utan called Susie had the honour of being the first of her species to reveal the genetic secrets of the species, quickly followed by ten wild individuals, five from Borneo and the rest from Sumatra.

The illustration below shows a tiny fraction of the three billion letters that form all the instructions that are needed to make an orang-utan. As we will see in Chapter 5, the code of life is composed from only four letters (A, C, T and G) which represent four individual chemical compounds, known as bases, which carry all the information of almost every living thing on Earth. In the case of the orang-utan, it's the information contained within this code that creates every component of these creatures' bodies, from their distinctive red hair to their long arms and short, bent legs.

The instructions to produce an orang-utan have remained consistent for millions of years, faithfully reproduced from one generation to the next. To do this, the orang-utan – and, indeed, all life on Earth – relies on one of DNA's most remarkable properties: its incredible stability and resistance to change.

To understand how this molecule is able to encode precious order from generations past, and pass it on so faithfully to the next, we need to understand the mechanism by which the replication of DNA occurs – a process that occurs every time a cell divides. From the moment of conception, the single cell of a fertilised orang-utan egg must divide trillions of times to produce the complex physiology and anatomy of a

**DNA CODON TABLE:** Standard genetic code

| 1st base | | 2nd base | | | | | | | | | 3rd base |
|---|---|---|---|---|---|---|---|---|---|---|---|
| | | **U** | | **C** | | **A** | | **G** | | | |
| **T** | TTT | (Phe/F) Phenylalanine | TCT | (Ser/S) Serine | TAT | (Tyr/Y) Tyrosine | TGT | (Cys/C) Cysteine | | | T |
| | TTC | | TTC | | TAC | | TGC | | | | C |
| **C** | TTA | (Leu/L) Leucine | TCA | | TAA | Stop (Ochre) | TGA | Stop (Opal) | | | A |
| | TTG | | TCG | | TAG | Stop (Amber) | TGG | (Trp/W) Tryptophan | | | G |
| | CTT | | CCT | (Pro/P) Proline | CAT | (His/H) Histidine | CGT | (Arg/R) Arginine | | | T |
| | CTC | | CCC | | CAC | | CGC | | | | C |
| | CTA | | CCA | | CAA | (Gln/Q) Glutamine | CGA | | | | A |
| | CTG | | CCG | | CAG | | CGG | | | | G |
| **A** | ATT | (Ile/I) Isoleucine | ACT | (Thr/T) Threonine | AAT | (Asn/N) Asparagine | AGT | (Ser/S) Serine | | | T |
| | ATC | | ACC | | AAC | | AGC | | | | C |
| | ATA | | ACA | | AAA | (Lys/K) Lysine | AGA | (Arg/R) Arginine | | | A |
| | ATG | (Met/M) Methionine | ACG | | AAG | | AGG | | | | G |
| **G** | GTT | (Val/V) Valine | GCT | (Ala/A) Alanine | GAT | (Asp/D) Aspartic acid | GGT | (Gly/G) Glycine | | | T |
| | GTC | | GCC | | GAC | | GGC | | | | C |
| | GTA | | GCA | | GAA | (Glu/E) Glutamic acid | GGA | | | | A |
| | GTC | | GCG | | GAG | | GGG | | | | G |

nonpolar    polar    basic    acidic    (stop codon)

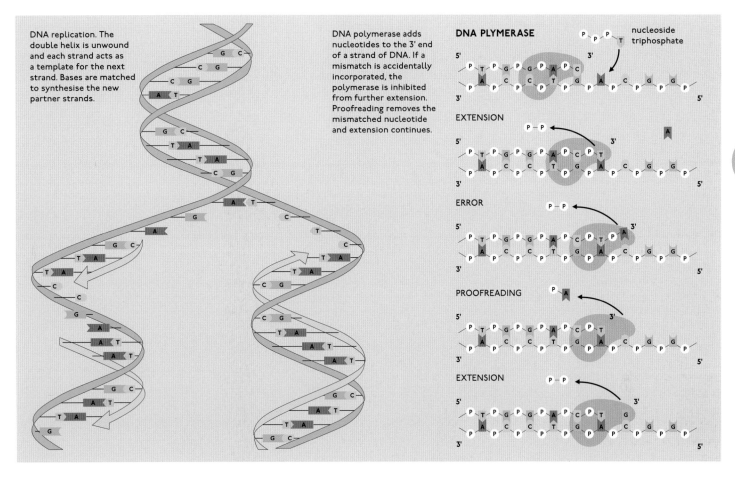

DNA replication. The double helix is unwound and each strand acts as a template for the next strand. Bases are matched to synthesise the new partner strands.

DNA polymerase adds nucleotides to the 3' end of a strand of DNA. If a mismatch is accidentally incorporated, the polymerase is inhibited from further extension. Proofreading removes the mismatched nucleotide and extension continues.

DNA PLYMERASE

nucleoside triphosphate

EXTENSION

ERROR

PROOFREADING

EXTENSION

113

*The instructions to produce an orang-utan have remained consistent for millions of years, faithfully reproduced from one generation to the next. To do this, the orang-utan relies on one of DNA's most remarkable properties: its incredible resistance to change.*

new life. Every time one of these cells divides, its DNA must be copied and, despite the gargantuan task of copying 3 billion letters, this process is highly resistant to copying errors. Led by the enzyme DNA polymerase, the chemical machine that does the copying is incredibly accurate, on average making only one mistake in a billion letters. To put that into context, it is like copying out each of the 775,000 words in the Bible around 280

times, and making just one mistake. By the time an individual produces the sperm or egg that will carry the code into the next generation, it will have copied itself trillions and trillions of times. It is this remarkable fidelity, combined with the steadying hand of natural selection, that allows advantageous adaptations to sustain, while detrimental adaptations most often quickly die before an animal has the chance to reproduce. So, natural selection acts to stabilise the code, ensuring that changes, if they do occur, are more likely to be beneficial, so that, over unimaginable periods of time, the blueprint changes to form a new species. It is this slow, guarded process that has enabled the tree of life to develop so broadly, but the deep-rooted ability of the system to conserve useful adaptations also means that, for most life on our planet, the vast majority of the code barely changes at all – if it works, life sticks with it and makes use of it in all manner of different forms.

Even though we're separated from orang-utans by more than 4 million years of evolution, what is really striking is just how similar we are. Orang-utans are one of the most human of animals; they share many behavioural traits that we would often define as being uniquely human. They pass on information, they teach their young, they nurture them for eight years before they let them go off on their own into the forest, and they teach them how to live. They learn which fruits are safe to eat, and which are poisonous; they learn which branches will hold their weight, and which won't; they learn how o build shelters from the rain. And they can do all this

| 16 M | 8 M | 4 M | 2 M | I M | 500,000 | 250,000 | 125,000 | TODAY |

H. neanderthalensis

→ H. sapiens

Genus

H. erectus

Family

→ Chimpanzee

→ Orang-utan

114

**PREVIOUS SPREAD:** Like all life on Earth, the orang-utan, shown here at the Sepilok Forest Reserve in Borneo, relies on DNA's incredible stability and resistance to change.

**BELOW:** Orang-utans share many behavioural characteristics with humans. They protect, nurture, and teach their young, and, most significantly, they are able to learn from memory.

**RIGHT:** The genes of a human being are 99 per cent the same as those of chimpanzees and bonobos.

115

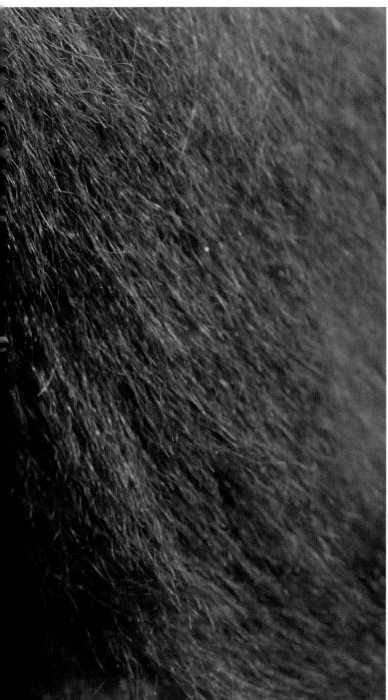

*All life on Earth is related – connected through our genetic code. DNA is the blueprint of life, and the keeper of a great story.*

because they have memory. They can remember things that happened to them in their life, they can learn from their memories, and they can pass them on from generation to generation. In the same way the DNA of these creatures enables a species to remember every single change that has ensured their survival, stretching back all the way through their ancestry, linking not only all primates but also all life on Earth.

Draw a tree of life of the primates and the first common ancestor we get to is with the chimpanzee and bonobos, about 4 to 6 million years ago – a blink of the eye in evolutionary terms. This means that if we compare our genetic sequence with that of a chimpanzee we find that our genes are 99 per cent the same. If we go back further to the split with gorillas, which occurred about 6 to 8 million years ago, we find that we share 98.4 per cent of our genes. Back in time again to the common ancestor we share with the orang-utan, and we find our genetic codes are 97.4 per cent the same. Even if we travel yet further back, to the split with the so-called old-world monkeys – animals such as macaques – we still find that our genetic codes are 94.9 per cent the same. In fact this journey can continue all the way back in time through our common ancestors with birds, reptiles, insects and bacteria, and no matter how seemingly unrelated to us these creatures become, we will still find sequences in the genetic code that are identical to sequences in our human cells. It reveals to us, with the most direct evidence, that all life on Earth is related – connected through our genetic code.

DNA is the blueprint of life, but it is also the keeper of a great story, perhaps the most astonishing story ever told, because our DNA not only connects us to every plant and animal alive today, but to every single thing that has ever lived. ◉

# CHAPTER 3

## SIZE MATTERS

## LIFE-SIZE DIFFERENCES

The smallest living thing on Earth is a thousand million times smaller than the largest. The tallest trees reach heights of over 100 m, and weigh more than 1,000 tonnes. The smallest bacterial cells are less than a millionth of a metre in length, and weigh a million millionths of a gram. From its beginnings on our planet about 3.8 billion years ago, life has branched, flourished and diversified into countless niches, creating a bewildering array of structure, form and function, from the simple yet ubiquitous bacteria to rare multicellular giants like the 200-tonne blue whales that roam our oceans with effortless grace. These life forms are radically different in all but their most basic biochemistry; they share the same planet, but they inhabit different worlds.

**ABOVE:** An African elephant and a male impala drink side by side from a waterhole in Chobe National Park, Botswana. There is enormous variation in the size, structure and form of the Earth's animals – but all are constrained by the laws of physics.

More than any other physical characteristic, it is size that dictates a living thing's relationship with the world, and there is a reason for this. The size, structure and form of living things are constrained by the laws of nature, and these are unavoidable, notwithstanding the undirected ingenuity of evolution. Even the blind watchmaker canna' change the laws of physics. What, then, is the smallest possible living thing, and what is the largest? Which laws limit the size of an organism, and what constraints are placed on evolution by those laws? And if organisms push towards these non-negotiable physical limits, what compromises are they forced to make? ◉

WONDERS OF LIFE

# SAME PLANET,
# DIFFERENT WORLD

The largest organisms that have ever lived on Earth are trees. The largest trees on Earth today – the costal redwoods, or *Sequoia*, of California – stand over 150 m tall, and occupy a volume of 1500 m³. Trees dwarf even the dinosaurs by any measure of physical stature we care to choose. Giant trees are found all over the planet, and Earth's smallest continent is no exception. Australia is home to the Tasmanian blue gum (*Eucalyptus globulus*), The Australian oak (*E. obliqua*) and the manna gum (*E. viminalis*), to name some of its grandest specimens, but the largest of them all is the mountain ash. *Eucalyptus regnans*, to offer its full name, is the tallest tree in Australia and the largest flowering plant in the world. Found in the wet highland regions of the southeastern state of Victoria and on the island of Tasmania, the mountain ash groves are almost Tolkienesque. In February, the shifting mists fight a running battle with the weakening late summer Sun, rendering every frame of towering columns in unique light, a fidgeting artistry calmed by the fragrant wet wood and fast-damped reverberation of the forest. This is a land of giants. The evergreens will grow a metre in a good year, reaching heights in excess of 100 m during their 400-year lives before physics intervenes like an irritating council official in a high-visibility vest to ensure that the world does not become too magical.

Righteously protecting conservative sensibilities in this particular case are the forces of gravity and electromagnetism, which together dictate the maximum size of a tree on Earth. The tree must be strong enough to support its own weight, and this depends on the nature of the building material. Wood is composed primarily of a long-chain carbon molecule called lignin, which we describe in detail in Chapter 5. Lignin's strength is determined by the strength of the chemical bonds that hold it together, which in turn are set by the strength of the electromagnetic force itself – a fundamental property of the Universe. There are stronger and lighter building materials in existence, of course; steel or carbon fibre, for example. But the tree has to be able to make lignin out of the raw materials available to it, using the biochemical processes of life. Lignin is one of the strongest and lightest materials available that can be easily assembled by living things, and it is the trade-off between the mass and strength of lignin that provides one of the constraints limiting the physical size of trees.

**LEFT:** *Eucalyptus regnans* (mountain ash) is a common sight in Australia and is one of the world's tallest trees. All trees, no matter what their size, need to be strong enough to support their own weight.

A tree must also be able to transport water, against the pull of gravity, all the way from its roots to its highest leaves. Trees raise water by capillary action, which depends on the way in which water molecules interact with each other, and the sides of the capillary tubes inside the tree, which are known as xylem. The interactions between water molecules are dominated by hydrogen bonds, as we saw in Chapter 1, and the strength of these is again ultimately down to the strength of the electromagnetic force. Trees of the scale of the mountain ash raise around 4 tonnes of water every day against the pull of gravity through capillary action, and so the details of these inter-molecular interactions play a crucial role in limiting their maximum height.

On a planet with a weaker force of gravity, trees constructed from lignin could therefore grow much taller than they do on Earth, because the strength of the electromagnetic force remains the same from planet to planet, whereas the weight of the building materials does not. This principle is illustrated by the size of mountains across the solar system. Olympus Mons, the tallest mountain on Mars, is almost three times the height of Mount Everest. On Earth,

it would weigh two and a half times as much as it does on Mars, and Earth's crust would be unable to support it. Everest is about as massive as a mountain can be on Earth. Mauna Kea, the Hawaiian volcano, is taller than Everest as measured from its base on the ocean floor, and it is gradually sinking under its own weight. What applies to mountains also applies to trees, and so a Martian forest might be expected to support towering giants that would reach heights of over 300 m.

This may sound rather whimsical, but the principle behind it is not. From the mountain ash of Australia to the billions of bacteria that cover its bark, all life is shaped and constrained by the universal laws of nature, and it is these laws that ultimately dictate the landscape of possible forms of life on Earth. The diversity in the sizes and forms of living things today reflects the complex interplay between the laws of physics and the vast array of niches and living spaces available for organisms to fill and exploit, stretching into virtually every corner and crevice of our planet.

We chose Australia as the backdrop to explore the story of size because it supports a range of habitats filled with a spectacular and, one must admit, photogenic and adrenaline-raising array of living things. We are making television, after all. With more species of reptile than anywhere else on Earth, some of the most venomous snakes in the world, and a large enough variety of spiders to disrupt the sleep of even the mildest arachnophobe, this is a country full of beautiful and simultaneously downright threatening animals. Most of the nasty ones are small, however, because in common with many regions of the world, the large land-based predators have been hunted to extinction. The marsupial lion, Australia's largest meat-eating mammal, died out 50,000 years ago shortly after the arrival of the first humans. To find giant, fast-moving predators today, therefore, we have to move off the dry land into Australia's rich coastal waters. ◉

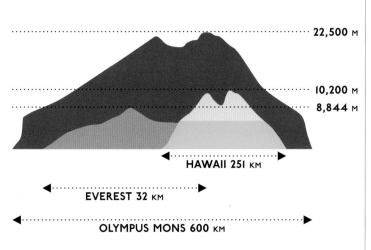

**MOUNTAINS ON MARS:** The towering mountains on Mars have been formed by vigorous volcanic activity; its largest, Olympus Mons, dwarfs the largest volcano on Earth.

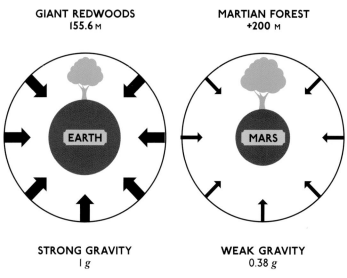

**TREES ON EARTH:** The gravitational force on Mars is equivalent to about 38 per cent of Earth's gravity, or $0.38\,g$, so an object exerting a force of 100 N on Earth would exert just 38 N on Mars.

# OCEAN GIANTS

**BELOW & BOTTOM:** Cage-diving for great white sharks, off the South Neptune Islands, is an awe-inspiring experience. Although great whites do not directly target humans, few people would feel comfortable diving among them without the protection of a cage.

**RIGHT:** The great white shark is highly adapted to its role as predator. It often approaches its prey from underneath and at high speed, breaking the surface with great force.

The South Neptune Islands sit in the deep waters at the mouth of Encounter Bay, close to the city of Adelaide. Their relative proximity to the coast belies their isolated, unprotected demeanour, enhanced by the swell rolling in from the Southern Ocean. On an overcast day, they feel almost Scottish. The islands are home to a lighthouse, some Australian sea lions and a colony of New Zealand fur seals, which are also inadvertently and doubtless reluctantly responsible for the island's only claim to fame. They serve as year-round food for a population of great white sharks (*Carcharodon carcharias*).

The great white is of course one of the world's iconic predators, and justifiably so. Reaching up to 6 m in length and weighing over 2,000 kg, it has been calculated that this creature bites with a force three times that exerted by the jaws of a fully grown African lion. They are rare and beautiful animals, and it was a privilege to be able to dive with them.

Although great whites do not directly target humans, they are by some way the most dangerous shark, and very few people dive with them unprotected. We filmed in a cage, suspended off the back of our dive boat. For me, this removed all sense of fear, but left an overwhelming sense of awe appropriately intact. The shape of an approaching great white

*Reaching up to 6 m in length and weighing over 2,000 kg, the great white shark can achieve speeds of over 32 km/h ... launching itself through the water in search of prey with exceptional velocity.*

is unmistakeable and graceful, gentle even. The animals of the ocean glide with dignified and unhurried ease, in contrast to the alarmist twitching of the inhabitants of the land. The limpid aquarium beyond the bars felt inviting, and I am sure I could have safely floated out into the unprotected blue. This impulse was tempered somewhat by the dive master, who in true Aussie style told me that it's easy to walk past the window of a cake shop when there is glass in it, but if the glass isn't there, then the temptation may prove too great. 'Its jaws are a metre wide, and they can swallow a man whole,' he said with the grin of a sledging fast bowler at the Gabba. The Merv Hughes of diving had a point, though.

Great whites are armed with an array of adaptations that make them extremely efficient predators. They display a multi-rowed arsenal of continuously growing serrated teeth, and around two-thirds of their brain is taken up by the olfactory lobes, allowing them to detect as little as one part per million of blood in the water. Great whites often approach their prey from below and at high speed; at 32 km/h they can break the surface with such ferocity that their not inconsiderable bulk is launched clean into the air. The ability to travel through water at high speed is clearly advantageous for a predator, but it is a difficult engineering problem that imposes significant physical constraints on the shape of sharks and indeed all high-speed marine animals. ◉

# THE PHYSICS
# OF A KILLER

**ABOVE:** A shark's gills have an extremely large, corrugated surface, enabling maximum contact with oxygen in the water.

Natural selection has crafted the great white shark for speed, and if choice were the right word, it didn't have any. The physics of fluids is universal and independent of biology, and so therefore are the engineering solutions that enable efficient high-speed motion through a fluid. There are two primary issues with moving though fluids. Firstly, the fluid itself has to be pushed out of the way as a shape moves through it. This becomes increasingly difficult as the cross-sectional area of the shape increases, simply because there is more fluid to move. Similarly, the force required to shift a given volume of water will increase with the density of the fluid. The volume of water moved each second is dependent on the velocity; if you move faster, you have to shift more water out of the way. The viscosity of the fluid will also play an important role. Viscosity is a measure of a fluid's resistance to flow, and is ultimately down to the inter-molecular forces between the liquid molecules themselves. Golden syrup is more difficult to swim through

**BELOW:** Near-invisible collagen structures called denticles (shown at 28x magnification) cover the shark's skin, reducing the Reynolds number by smoothing the flow of water around its body.

where ρ is the density, V is the velocity, μ is the (dynamic) viscosity and D is a quantity related to the cross-sectional area known as the hydraulic diameter. The Reynolds number is a useful quantity because it quantifies how efficiently a particular shape will move through a given fluid. The higher the Reynolds number, the greater the influence of drag and so the less efficiently the shape moves through the water. The selection pressure on a great white shark, therefore, might be expressed in terms of the advantage conferred by body shapes that reduce the Reynolds number. This leads to a characteristic geometry, which is also found in torpedoes and submarines. The maximum width is around a third of the way down its body, and the maximum width of its body is around a quarter of its length.

Despite this fine-tuning, however, the Reynolds number of a great white is still quite high, primarily because the shark is big and is heavily penalised by the need to push a large amount of dense water out of the way as it swims. The consequence of a high Reynolds number is that the flow of fluid across the surface of the shark's body is not what physicists call laminar flow – it is not smooth and streamlined, but turbulent, and this increases as the animal speeds, twists and turns in the chase, an effect that is visible in the chaotic motion of the water left in its wake.

Clearly, a high Reynolds number is a problem, and any adaptation that acts to lower it will be positively selected for in the population. This is the reason for the shark's quite fascinating skin, which acts to reduce the Reynolds number by smoothing the flow of water around its body. Sharks are covered in scales called dermal denticles, near-invisible collagen structures made of the same material as their teeth. They are aligned parallel to the flow of water, and are ribbed with longitudinal grooves, making the surface of the shark more streamlined as it moves through the water. Recent research at the University of Alabama suggests that the denticles may increase the shark's efficiency through another mechanism. The scales are loosely embedded in the skin, tethered with rubber-band-like tendons, allowing each one to move independently. It is thought that this allows them to act in a similar fashion to the dimples on a golf ball, affecting the wake behind the shark and reducing a phenomenon known as pressure drag. The denticles differ in size and flexibility over the shark's body, having the most flexibility in those areas that create the most drag, such as behind the gills.

Biology is obviously constrained by the laws of nature; otherwise living things would be supernatural! But there are cases in which the laws of physics apply such stringent constraints that, through natural selection, they determine to a large extent the form of the animal. The great white is an excellent example. For a high-speed marine predator to be so large, it must be shaped like a shark because the laws of fluid dynamics dictate it. Sharks with different body shapes would be slower, or expend significantly more energy reaching high speeds, and would therefore not have been so successful; gradually, from generation to generation, natural selection honed the shape of the shark. ◉

than water, partly because it is denser, but also because it is sticky. There is a dimensionless physical quantity (a pure number, in other words), known as the Reynolds number, which is widely used in the design of aircraft and submarines, and indeed in many problems that involve the flow of gases or liquids around shapes. It is the ratio of the inertial forces on a shape – the difficulty of shifting the fluid out of the way – to the viscous forces – the stickiness. If you think a little, you could probably write down the formula for the Reynolds number and get it qualitatively correct. It must depend in some way on the density of the fluid, the velocity of the shape through the fluid, the cross-sectional area of the shape, and the viscosity of the fluid. If you also want all the measurement units to cancel out, then there's not much it could be other than

$$Re = \frac{\rho V D}{\mu}$$

## SIZEABLE GIANT

A great white shark can reach 6 m in length and over 2,000 kg in weight. At this size it could exert a bite force of over 18,000 Newtons. Reaching maturity at around 15 years of age, it can have a life-span of over 30 years.

## PERFECT PROPORTIONS

Water is 800 times denser than air so everything about the shape of a great white shark is streamlined to achieve supreme maneuverability and stealth through the water. Humans can swim (at a dead sprint) up to about 10 km/h. Great white sharks can swim at about 32 km/h

Can exceed 6 m long

Its maximum width is around a third of the way down its body

## STREAMLINED SCALES

Shark skin is much tougher than whale skin and has actually been used as sandpaper. It's made up of millions of miniature teeth, called dermal denticles, with three longitudinal ridges that improve the shark's performance when it swims. They also provide an extremely tough outer coating against shark bites.

0.1mm

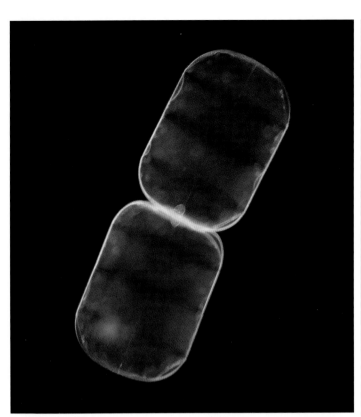

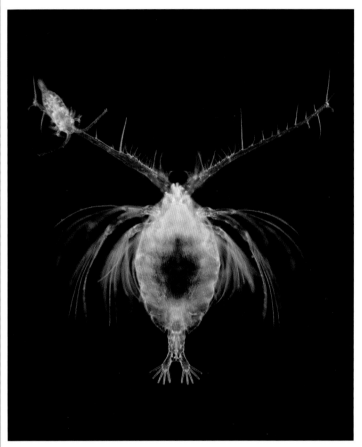

**LEFT:** Plankton come in all shapes and sizes. This *Ethmodiscus* belongs to the phytoplanktonic group of algae and has exceptional buoyancy.

**BELOW LEFT:** Amphipods live in almost all aquatic environments. They float with the currents, and, like most species of plankton, their body shapes are not streamlined.

**BELOW:** *Pontellina plumata* (marine copepod), with a smaller copepod accidentally trapped on one of its antennae. Plankton such as these look totally unlike the inhabitants of our larger world.

**RIGHT:** *Conochilus unicornis*. The density of water is almost irrelevant for plankton such as this, because they have to move so little of it out of their way.

# SMALL IS BEAUTIFUL

The physical constraints that determine the form of animals in the oceans change dramatically with size. Take a look at the creatures on these pages. Our oceans are teeming with these miniature organisms, collectively known as plankton. These organisms are not usually classified according to their phylum, class, order or family. Rather, they are classified according to their place of residence. In the case of plankton, this is the pelagic zone of our oceans – the area that is neither close to the bottom nor near the shore. Any organism, be it animal, plant, bacteria or archaea, can be classified as plankton, which might be seen as a term describing their most important common characteristic.

The word 'plankton' comes from the Greek planktos, meaning errant. These are organisms that float with the currents, drifting passively along the shifting highways that carry life around the seas. Any animal that can resist this flow falls into a different category, known as nekton – creatures such as fish and marine mammals that have the anatomical modifications and strength to exert full control over their position.

Although plankton are unable to control their horizontal position, they can and do adjust their vertical position. They often swim hundreds of metres in a single day as they migrate up and down a vertical column of water. As a result, plankton are responsible for the largest daily migration of biomass on our planet.

Organisms don't have to be microscopic to be classified as plankton; some of the smaller jellyfish and other cephalopods are known as megaplankton, and are visible to the naked eye. But although being microscopic isn't a

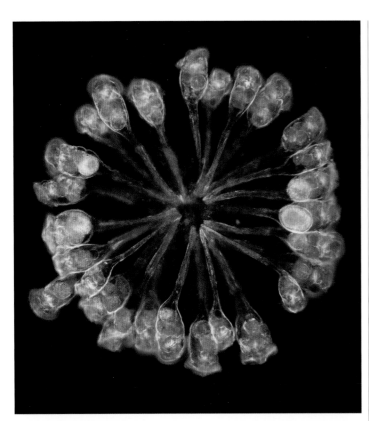

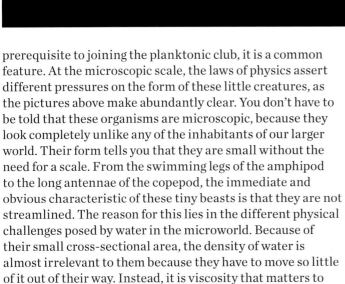

**BELOW LEFT:** This sea butterfly, or pteropod, has two large, wing-like parapodia that it uses for swimming. Rather than attempting to adjust their horizontal position, plankton swim up and down a vertical column of water.

**BELOW:** Plankton such as this waterflea (*Ceriodaphnia reticulata*) live in a world of small Reynolds numbers.

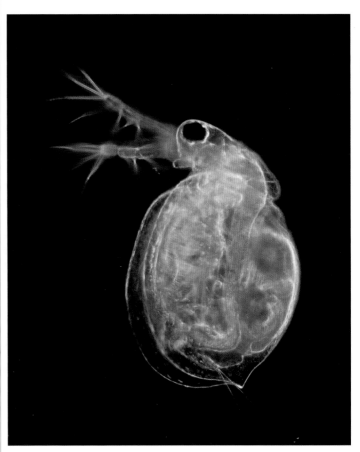

prerequisite to joining the planktonic club, it is a common feature. At the microscopic scale, the laws of physics assert different pressures on the form of these little creatures, as the pictures above make abundantly clear. You don't have to be told that these organisms are microscopic, because they look completely unlike any of the inhabitants of our larger world. Their form tells you that they are small without the need for a scale. From the swimming legs of the amphipod to the long antennae of the copepod, the immediate and obvious characteristic of these tiny beasts is that they are not streamlined. The reason for this lies in the different physical challenges posed by water in the microworld. Because of their small cross-sectional area, the density of water is almost irrelevant to them because they have to move so little of it out of their way. Instead, it is viscosity that matters to microscopic plankton, and this dictates how evolution has shaped and driven their form. If you are small, water feels like thick, gloopy treacle would to us. They live in a world of small Reynolds numbers, where sticky forces present the largest challenge to getting around. In order to move, evolution has devised retractable paddles and oars which allow them to climb through the water. Moving millimetres at a time, there is no effortless gliding for these creatures; when they stop paddling they simply stop moving.

Here, then, is a contrasting example of an animal's size and environment dictating its form. For animals the size of plankton it is the viscosity of water that matters. Water is sticky on small scales, and so these animals have evolved to climb their way through it. Large animals, on the other hand, have to contend with the mass of water they have to force out of the way simply to move around, and the stickiness plays a subsidiary role. Animals of different sizes can live in the same environment, but because of the different physical challenges they face, their physical forms can be radically different. ◉

# OF ROYAL CROWNS AND OCEAN GIANTS

The crown and the gold have equal weight

The crown displaced more water than the gold

**ARCHIMEDES' CROWN EXPERIMENT**
After measuring the water displaced by the crown, Archimedes dropped into the same jar a lump of gold the same weight as the crown. This displaced less water than the crown, demonstrating that the crown was not pure gold.

Whether he ran naked from the bath shouting Eureka or not, Archimedes' eponymous principle explains why giants can exist in the oceans. Born in the costal town of Syracuse in 287 BC, Archimedes would have been familiar with great marine mammals. Dolphins regularly migrate through the region, and bottlenose whales are often glimpsed off the coast of Sicily. According to legend, Archimedes' interest in the behaviour of objects submerged in water was sparked by a problem put before him by King Hiero II, the Greek Sicilian King of Syracuse. Hiero commissioned a ceremonial crown to be made by a local goldsmith, and provided pure gold for the manufacture. On the crown's completion, however, the King suspected that he had been cheated, believing the gold to have been mixed with cheaper metals. Archimedes was asked to prove the deception without damaging the crown. Archimedes knew that if the gold had been mixed with metals such as silver, then its density would have decreased accordingly. But measuring the density requires a precise determination of the volume, and without melting the crown down and recasting it into an easily measurable shape, this seemed impossible. Legend has it that Archimedes sat down in a nice hot bath to consider the problem and, during this auspicious soak, noticed that the level of the water rose as he sank into the water. He immediately realised that the volume of any body could be calculated by measuring the volume of water it displaces when fully submerged.

With the volume determined, he could measure the mass of the crown and calculate its density, which is simply the mass per unit volume. If this turned out to be less than the density of pure gold, then the jeweller would be, as we say in Manchester, in the shite.

## ARCHIMEDES' PRINCIPLE

Archimedes' principle states that the upward buoyancy force on an object immersed in a fluid is equal to the weight of the displaced fluid. Here, an aluminium block is first weighed in air (78 g) with a spring scale. Then the block is immersed in a beaker filled with water and the measured weight is reduced to 52 g, due to the buoyancy force.

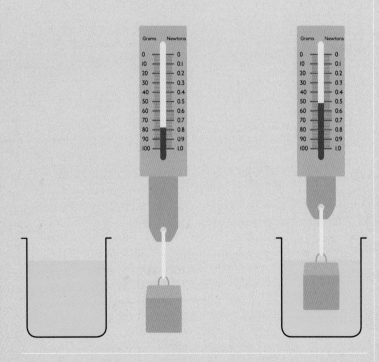

**LEFT:** According to legend, Archimedes discovered his eponymous principle while sitting in the bath.

**TOP:** Mass and weight are different concepts, as shown by this experiment with water and scales.

Whether these events actually took place is immaterial, and in fact the determination of the volume of an object by immersing it in water is also not particularly fundamental. But there is a deeper physical insight lurking here, which he described in his work 'On floating bodies'; 'any object, wholly or partially immersed in a fluid, is buoyed by a force equal to the weight of the fluid displaced by the object'. This is a rather advanced observation for its time, coming as it did almost two millennia before Newton's Principia. It means that, if a body is less dense than water, it will experience an upward force when submerged, equal to the weight of an equivalent volume of water minus the weight of the body. This force will push it up to the surface. Similarly, if a body is denser than water, it will sink, although its apparent weight (by which we mean the force acting downwards on the body) will be reduced by an amount equal to the weight of an equivalent volume of water. This is why heavy things sink much more slowly than they fall in air. This effect is of overriding importance for animals living below the waves, because it effectively means that they are weightless. At the risk of labouring the point, all animals are made up predominantly of saltwater, which means that they weigh very close to an equivalent volume of saltwater! This means that their apparent weight is always very close to zero, and in fact many marine animals use organs such as swim bladders to maintain neutral buoyancy, which is equivalent to arranging things such that your apparent weight is precisely zero. Being weightless means that you do not have to deal with the inconvenient force of gravity, and this is the reason why the largest animals that have ever lived, blue whales (*Balaenoptera musculus*), are to be found in the oceans. ◉

# BIG THINGS DON'T JUMP

I n his magnificent history of science in the romantic age, Richard Holmes asserts that 'The Age of Wonder' began in 1768 with Captain James Cook's round-the-world voyage aboard HMS *Endeavour*. The voyage was commissioned jointly by the Royal Navy and the Royal Society to explore the South Pacific Ocean, and was timed to coincide with one of the rarest of predictable astronomical events, the transit of Venus across the face of the Sun on 3–4 June 1769. Few celestial events were of comparable scientific importance to the curious minds at the Royal Society, because by simple triangulation, observations of the transit, timed and repeated around the globe, allow for a precise measurement of the distance between the Earth and the Sun. The crew of the *Endeavour* were to observe the transit from Tahiti. In common with many of the great scientific voyages, the reasons for funding the expensive expedition were partly political. The Admiralty was interested in rumours of a great Southern continent, of which New Zealand was the northern promontory, and particularly keen to ensure that, if such riches indeed existed in the Southern Ocean, the French didn't get there first. One gets the impression that the scientists and sailors aboard the *Endeavour* didn't believe a word of this, but in a noble tradition that continues to this day, they put their scepticism to one side and took the politicians' cash in the name of science.

**ABOVE:** The naturalist Joseph Banks was astonished to encounter a hopping animal as large as a kangaroo, as illustrated in this engraving by John Hawkesworth.

**ABOVE RIGHT:** Kangaroos are the only large animals that hop in order to propel themselves forward.

**OVERLEAF:** The kangaroo is a very efficient hopper; its energy consumption decreases once it exceeds 10 km/h, and doesn't rise again until cruising speed of about 40 km/h is reached.

Aboard the ship were a number of scientists including the astronomer Charles Green, naturalist Daniel Solander and one of the towering figures of British science, Joseph Banks. Banks was a rich man, and funded his own 8-man natural history department aboard the *Endeavour* at a cost of £10,000, a very large sum indeed in 1768. Ever the Yorkshire squire, he also took a pair of greyhounds. Holmes describes Banks as being 'tall and well built, with an appealing bramble of dark curls', and of possessing the 'dreaming inwardness of Romanticism'. A charismatic figure, in other words, who went on to become the longest-serving president of the Royal Society, from 1778 until his death in 1820. Among Banks' great achievements was the founding of the Royal Institution

of Great Britain in 1799, with a view to promoting the idea, which is self-evidently correct, that science and engineering are the foundation upon which the economy of a country, or in Banks' time the economy of a growing empire, thrives for the benefit of all its people. Humphry Davy and his protégée Michael Faraday are among the figures who found their scientific home at the Royal Institution and transformed the modern world. The prospectus for the funding of the Royal Institution is a document that is as relevant today as it was over 200 years ago. 'But in estimating the probable usefulness of this Institution, we must not forget the public advantages that will be derived from the general diffusion of a spirit of experimental investigation and improvement among the higher ranks of society. When the rich shall take pleasure in contemplating and encouraging such mechanical improvements as are really useful, good taste, with its inseparable companion, good morals, will revive: – rational economy will become fashionable: – industry and ingenuity will be honoured and rewarded; and the pursuits of all the various classes of society will then tend to promote the public prosperity.' In today's language, substitute 'political class' for 'rich' and post this off to Number 10!

The transit observations of 1769 were a great success; the results were collated and published in the Philosophical Transactions of the Royal Society in 1771, and the mean distance from the Earth to the Sun was calculated to be

93,726,900 miles. This is within 1 per cent of the current value, determined by radar.

Sailing onwards around the world, the expedition reached the Queensland coast in the early summer of 1770, where it was stranded for weeks after the *Endeavour* struck the Great Barrier Reef. Banks, Solander and the Finnish botanist, Herman Spöring, took the opportunity afforded by this unexpected layover to carry out the first detailed study of the flora and fauna of Australia, and in doing so introduced Europeans to now-familiar species such as eucalyptus. The star of the antipodean show, however, was a bizarre animal unlike anything Banks had seen before. After initial sightings by crew members, Banks wrote: 'In gathering plants today, I myself had the good fortune to see the beast so much talked of, tho but imperfectly; he was only like a greyhound in size and running, but had a long tail, as long as any greyhound's; what to liken him to I could not tell, nothing certainly that I have seen at all resembles him.' Cook recorded the name of the animal after hearing it spoken by the Aboriginal people who lived along the northeast coast; *gangurru*, which Cook translated as kangaroo'.

The kangaroo must have looked strange to say the least to the crew of the *Endeavour*. They look strange to me, hopping along the highways and through the towns of Australia just as squirrels and foxes do in urban England. Grey kangaroos

(*Macropus giganteus*) are common, but we travelled to the 'Capital of the Outback', Broken Hill, a dusty mining town rich in heritage and frontier architecture, to film Australia's largest surviving land mammal, the much less common red kangaroo (*Macropus rufus*). These animals, which can stand taller than an adult human, are the largest of around 50 species of Australian macropods, a family that includes wallabies and pademelons. Macropod means, appropriately for the red kangeroo, 'large foot'; the animal's gait is dominated by its hind legs, which lend it an ungainly appearance when ambling around the bush in search of food. Natural selection is not given to producing inefficient animals, however, and the reason for the kangaroo's apparently eccentric hindquarters becomes clear when they start to move fast. Kangaroos are the only large animals to have ever lived that use hopping as a means of propulsion; there is no evidence in the fossil record of any other similarly sized hopping animal. This is in some ways surprising, because hopping enables kangaroos to move very fast, and with great efficiency as their speed increases. Unlike animals with more conventional methods of locomotion, the kangaroos' energy consumption decreases as they exceed 10 km/h, and doesn't rise again until they exceed their cruising speed of 40 km/h. The reasons behind this efficiency can be found in the mechanical details of their anatomy.

**SCALING LAWS:** While the weight of a structure increases with the cube of its dimensions, the area of the load-bearing sections increases only with the square of its dimensions. Any structure scaled upwards on Earth will eventually fail under its own weight.

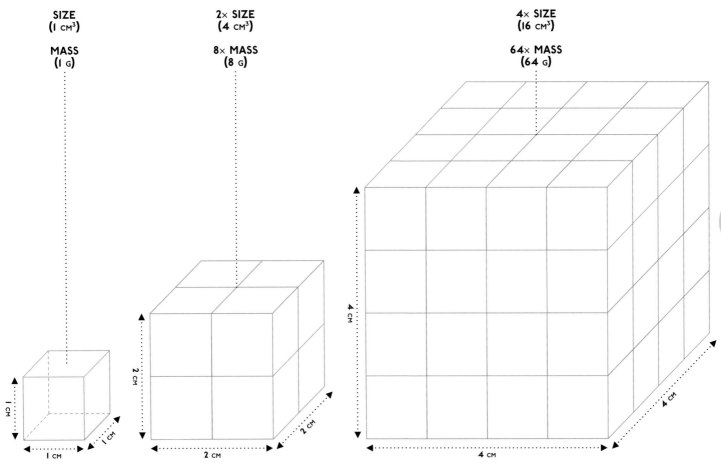

SIZE (1 cm³)

MASS (1 g)

2× SIZE (4 cm³)

8× MASS (8 g)

4× SIZE (16 cm³)

64× MASS (64 g)

The tendons in a kangaroo's legs, particularly its Achilles tendon and the tendons in the tail, store energy on impact and then release it like a tennis ball, powering the animal forward. This is an ingenious engineering solution not unlike the regenerative breaking technology in modern hybrid cars, which converts kinetic energy into stored potential energy which can be used to help drive the car forward. In the case of the kangaroo, a large proportion of the not-inconsiderable energy dissipated as its 90 kg body hits the ground is stored in its elastic tendons and then re-used, allowing it to leap up to 9 m in a single bound. This seems such a sensible thing to do that, at first sight, it appears to be a mystery why evolution has not deployed this technology in other, larger animals, but in this case the reasons are not biological. Giant hopping dinosaurs or elephants have never existed because of the physical constraints imposed by the strength and mass of the materials biology has at its disposal, because as animals get larger, gravity places increasing structural demands on their bodies.

## ON BEING THE RIGHT SIZE
In 1926 the British biologist J B S Haldane published a now-famous essay entitled 'On Being the Right Size', in which he considered how the size of animals affects their form and structure. Would it be possible for the giants of fairy tales to exist? Could it be possible for a hare to grow as large as a hippopotamus? The answer to both these questions is no. But why? What properties of the physical world set the scale of living things? Why can't every dimension of an animal, from the length of its legs to the the size of its heart, simply be scaled up to give a fully functional giant? The answer lies in the way that certain fundamental physical quantities vary with size. Consider the volume of a cube, for example. If its sides measure 2 cm in length, then its volume is 2 cm x 2 cm x 2 cm = 8 cm³. If we now double the size, so that the cube has sides of length 4 cm, then the volume is 4 cm x 4 cm x 4 cm = 64 cm³. Doubling the size of a cube therefore results in its volume increasing by a factor of 8. Double the dimensions of the cube again, and the volume increases to 512 cm³, another factor of 8. Mathematically, we can say that the volume of an object increases as the cube of its size. This is important for animals because their mass increases in proportion to their volume. This should be obvious, since the volume of something is a measure of the amount of 'stuff' inside it. If the size of an animal is doubled, then its mass will increase by a factor of 8, and this changes the structural demands on the animal's skeleton dramatically because the force of gravity acts in proportion to the mass. This principle is wonderfully demonstrated by the fossilised bones of Australia's long-extinct giants. ◉

# LOST GIANTS

**LEFT:** This artist's impression shows a procoptodon, which lived in Australia during the Pleistocene epoch, and is the largest kangaroo known to have existed, standing at over 2 m tall.

**BELOW:** Diprotodon, shown in this artist's impression, is believed to have lived in Australia between 1.9 million and 10,000 years ago. Its closest living relative is the wombat.

**BOTTOM:** As animals get bigger, an increasing amount of their body mass is made up of bone. Dromornis, illustrated here, is believed to be the largest bird that ever lived, standing 3 m tall.

137

For hundreds of thousands of years before the arrival of humans, Australia was populated by creatures that would have dwarfed their modern cousins. The dromornis bird, a sort of giant goose, stood 3 m tall and weighed half a tonne. Marsupial lions the size of African lions may have been the top predators, although they would have faced competition from the largest known terrestrial lizard, the 7 m-long *Varanus priscus*. Yet for sheer size, few creatures on the continent came close to the diprotodon. The first diprotodon skeleton was discovered in a cave in New South Wales in the early 1830s by Lieutenant Colonel Sir Thomas Mitchell, one of the great European explorers of the Australian wilderness. Mitchell sent the bones back to Britain to be examined by Sir Richard Owen, the founder of London's Natural History Museum and the man responsible for coining the term 'dinosaur'. But the diprotodon wasn't one of Owen's 'terrible lizards'; rather, it was the largest marsupial ever to have lived.

The size of a rhino, the diprotodon stood 2 m tall and 3 m from nose to tail, was covered in hair, and had two protruding front teeth, a huge pouch and powerful claws. It would have looked like a gigantic lumbering wombat. Diprotodon fossils have been found all over Australia, spanning a 25-million-year tenure during which a steady evolutionary drift towards gigantism resulted in this largest of all the Australian megafauna. Some 50,000 years ago, along with almost all the other Australian giants, the diprotodon disappeared, which corresponds roughly to the arrival of humans across the shallow seas and island chains that would have linked Australia to Eurasia after the last glacial maximum.

The closest living relatives of the diprotodon are wombats, and a comparison of their skeletons provides a vivid example of the demands of increasing size on the bones of land-based animals. The wombat – an animal around the size of a small dog – weighs 35 kg and is about 1 metre in length. The diprotodon was 3 m in length, and from our previous discussion of the relationship between size and mass, we might estimate that it would have been approximately 30 times as heavy as a wombat, weighing in at around a tonne. Diprotodons were in fact somewhat heavier than this, at 2 tonnes, which is near enough for jazz, and certainly biology.

Looking at the femur of a 2-tonne diprotodon, the bones look very similar in structure to the femur of a wombat, with the same anatomical features, as would be expected for closely related species. But the relative dimensions of the bones are very different. A typical adult wombat femur is 15 cm in length, whereas the diprotodon's is 75 cm; 5 times as long. The cross-sectional area of the wombat femur is around 2 cm$^2$; however, the diprotodon femur has a cross-sectional area of 80 cm$^2$, an increase of a factor of 40! This dramatic increase in bone size is necessary to support the increased mass of the animal, because the strength of a bone increases in proportion to its cross-sectional area. If the bones of the diprotodon were simply wombat bones scaled up in every dimension without any modification, they would snap when the animal began to move.

The photograph below shows a selection of femur bones from animals of different sizes, starting with one of the smallest marsupials in Australia – the marsupial mouse, or Antechinus. Next is a potoroo, a marsupial around the size of a rabbit, followed by a Tasmanian devil, a wombat, a dingo, the red kangaroo – the largest marsupial in Australia today – and the diprotodon. Finally, at the top of the picture, is a model of the femur of a rheotosaurus, a sauropod dinosaur 17 m long and weighing 20 tonnes.

As animals get bigger, it is unavoidable that an increasing amount of their body mass is made up from the thicker bones needed to support the increased bulk. This is ultimately the result of two scaling laws that are physical, not biological, in origin. As an animal increases in size, its mass increases as the cube of the increase in size. For animals on land, this delivers a steep increase in the forces the skeleton has to deal with. Bone strength increases as the square of the cross-sectional area of the bone, and so animals require thicker and

*It is not the availability of food or the outcomes of evolution that ultimately decide the size of the largest land-based animal – it is gravity.*

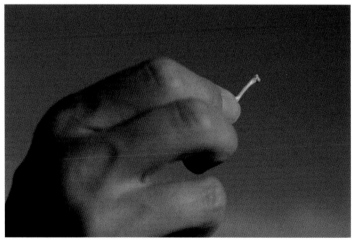

**LEFT:** Femurs come in vastly different sizes in different species.

**TOP:** While the femur of the 35 kg wombat is 15 cm in length, and that of the 2-tonne diprotodon is 75 cm (pictured here), the cross-sectional area of the latter is a remarkable 40 times that of the smaller bone.

**ABOVE:** The tiny femur of a marsupial mouse (Antechinus).

thicker bones as they get larger. This in turn increases the percentage of the animal's body mass that is made of bone, and since bone is relatively dense, this in turn increases the mass of the animal.

Ultimately, then, the size of animals on land is restricted by the strength of bone and the mass of the Earth. On a planet such as Mars, given the same bone strength, animals could be more massive because Mars's gravitational pull is around a third that of the Earth. This would permit, in principle, larger animals to roam the surface of the red planet.

The mass of our planet therefore sets an upper limit on the size of the animals that inhabit it. The great dinosaurs like diplodocus and argentinosaurus existed on the limit of bone strength, and would have been in extreme peril if they fell over. This would have determined their method of locomotion; tripping would have been avoided at all costs. This is why, despite the efficiency of its method of locomotion, we see no animal bigger than the majestic red kangaroo hopping on the surface of our planet. The force exerted by our planet on the bones of a hopping elephant would be too great, and an elephant with bones thick enough to withstand the stress of hopping would be too massive to launch itself off the ground in the first place.

Gravity plays an increasingly dominant role in the form and function of large animals. But at the other end of the mass scale, it is effectively irrelevant. ◉

# THE WORLD OF THE SMALL

There was a time in the Earth's history, around 300 million years ago, when giant insects roamed the planet. Dragonflies with wingspans the size of a hawk soared in the air, millipedes over a metre in length darted around, and cockroaches bigger than a human foot could crush roamed the land. The reasons why giant insects don't exist today is still the subject of scientific debate, but current thinking suggests that it was the oxygen-rich atmosphere that existed during the Paleozoic era that allowed giant insects to thrive.

**RIGHT:** Goliath beetles (*Goliathus orientalis*) are among the largest insects on Earth. The mounted specimen shown here is about 8.5 cm long.

Around 300 million years ago, our atmosphere was 35 per cent oxygen, rather than the 21 per cent today. Insects don't have lungs and don't transport oxygen around in the blood as we do. Instead, they rely on a system of tubes, called trachea, connected to holes in their bodies known as spiracles. Oxygen enters through these holes, and carbon dioxide is expelled. As the insect gets larger, the percentage of its body taken up by the tracheal system increases rapidly as more and more oxygen is required. Recall that an animal's volume, and therefore the amount of living tissue requiring oxygen, increases as the cube of its size. Insects in the Paleozoic era could afford a smaller tracheal system, relative to their volume, to deliver oxygen around their bodies, simply because the oxygen content of the gas passing through the tubes was higher. Calculations of the maximum size of beetles, based on the maximum possible size of the tracheal openings, suggest that a 21 per cent oxygen atmosphere can only support beetles of around 15 cm in length, and indeed, the largest known beetle on Earth today, the Titanic longhorn beetle (*Titamus giganteus*), is approximately this size.

Today, the giant insects are long gone, but insects still dominate the planet. Over a million different species have been discovered, but entomologists believe that there may be 10 million more awaiting discovery. This means that over 75 per cent of all known animal species are insects, and the true figure may well be significantly greater than 90 per cent. This dominance goes relatively unnoticed primarily because of the restriction on their size, which, as we discussed above, is probably due to non-biological geometric restrictions on their tracheal networks and the oxygen content of the atmosphere.

Perhaps it is a good thing that insects are small. It is estimated that there are over 10 billion, billion individual insects alive today, living in virtually every environment on the planet.

Of all the insect groups, it is the Coleoptera, commonly and collectively known as beetles, which boast the greatest number of species. Over 400,000 different species are known, but, as with all insects, there is no doubt that there are hundreds of thousands more yet to be discovered and classified. Beetles have fascinated naturalists for centuries. Darwin himself was a keen beetle collector, and it was his passion for collecting and documenting beetles that led to his first brush with fame. Long before he rose to prominence, he was mentioned for the first time in print in J F Stephens' *Illustrations of British Entomology*. Darwin, like thousands of others before and after, became captivated by the limitless extravagance in form and colour of these often bizarre creatures.

J B S Haldane, an evolutionary biologist who is credited with a major role in the development of neo-Darwinian thinking, is famous for the (possibly apocryphal) quotation:

*'...if one could conclude as to the nature of the creator from a study of creation, then it would appear that God has an inordinate fondness for stars and beetles...'*

Whether or not Haldane ever actually spoke these words, it is nevertheless a great quote, and it was aimed at theologians who claimed that the nature of God could be discerned from His creation – the implication being that man, being the greatest creation of all, is crafted in God's image. ◉

# INSIDE AN INSECT

Insects are some of the most recently evolved members of the arthropod family. The segmented bodies of arthropods evolved from worm-like ancestors. Each segment has its own pair of appendages, which have evolved to meet different needs, as is exemplified here in this illustration of a fly.

**FLYING**
Wings are thought to have evolved from the gills that were present when their ancestors lived in the water. Most winged insects have two pairs of wings; in flies the rear pair is reduced to form a stabilizer to help the insect orient itself. In beetles the front pair form hard protective structures called elytra.

**METAMORPHOSIS**
Insects either undergo full metamorphosis, following the stages egg – larva – pupa – adult, or they experience partial metamorphosis, following the stages egg – nymph – nymph – nymph – adult.

FOREWING

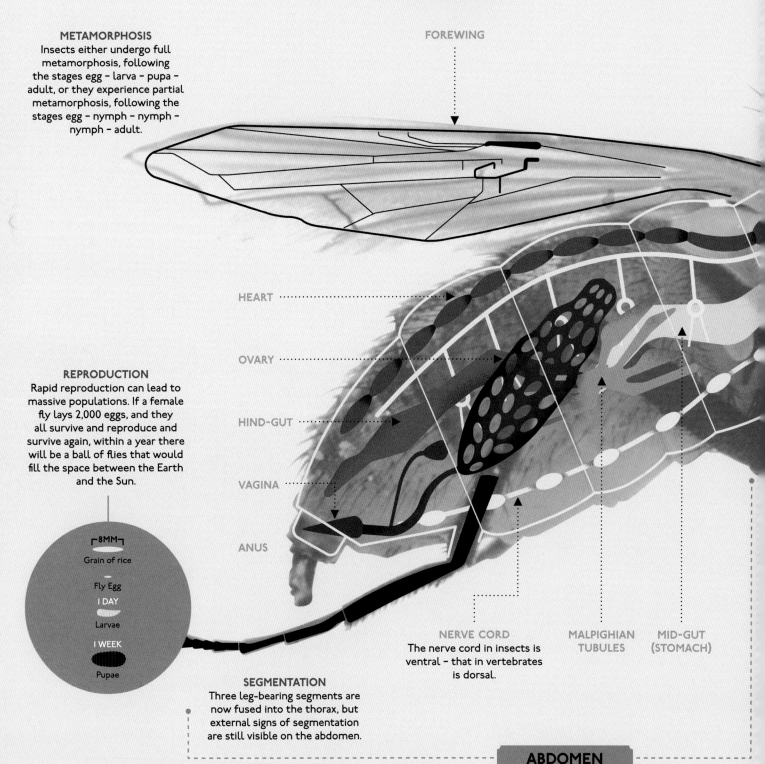

HEART

OVARY

HIND-GUT

VAGINA

ANUS

**REPRODUCTION**
Rapid reproduction can lead to massive populations. If a female fly lays 2,000 eggs, and they all survive and reproduce and survive again, within a year there will be a ball of flies that would fill the space between the Earth and the Sun.

⌐8MM⌐
Grain of rice

Fly Egg
I DAY

Larvae
I WEEK

Pupae

**SEGMENTATION**
Three leg-bearing segments are now fused into the thorax, but external signs of segmentation are still visible on the abdomen.

**NERVE CORD**
The nerve cord in insects is ventral – that in vertebrates is dorsal.

MALPIGHIAN TUBULES

MID-GUT (STOMACH)

ABDOMEN

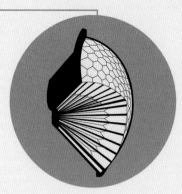

## BREATHING
Respiration takes place through the spiracles and is more or less passive. But insects can also pump air in and out by moving their abdomen.

## PROCESSING
The brain is good at processing visual signals – in particular movements and chemical signals associated with memory formation.

## 360 VISION
The fly has four thousand separate lenses in each eye – eight thousand in total – providing wide-angle vision which is in fact omni-directional.

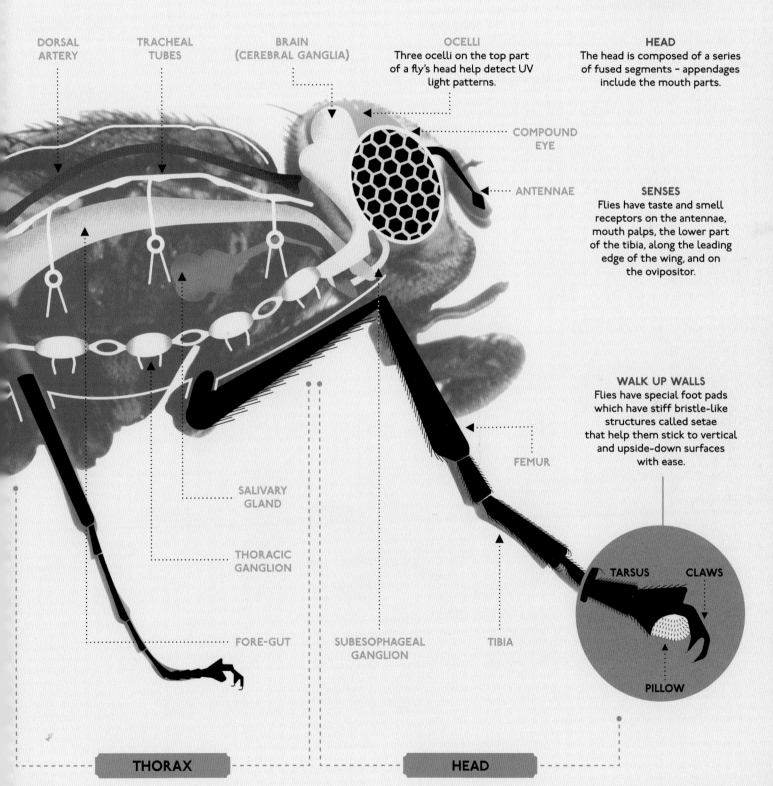

DORSAL ARTERY

TRACHEAL TUBES

BRAIN (CEREBRAL GANGLIA)

## OCELLI
Three ocelli on the top part of a fly's head help detect UV light patterns.

## HEAD
The head is composed of a series of fused segments – appendages include the mouth parts.

COMPOUND EYE

ANTENNAE

## SENSES
Flies have taste and smell receptors on the antennae, mouth palps, the lower part of the tibia, along the leading edge of the wing, and on the ovipositor.

## WALK UP WALLS
Flies have special foot pads which have stiff bristle-like structures called setae that help them stick to vertical and upside-down surfaces with ease.

SALIVARY GLAND

THORACIC GANGLION

FEMUR

TARSUS

CLAWS

FORE-GUT

SUBESOPHAGEAL GANGLION

TIBIA

PILLOW

**THORAX**

**HEAD**

# BEETLE MANIA

**BELOW:** This collection of butterflies, beetles and other insects demonstrates their extraordinary diversity of forms and colours.

**BELOW RIGHT:** The rhinoceros beetle proves its incredible strength.

I would hazard a guess that most scientists will be able to identify a moment, or perhaps a particular natural phenomenon, possibly obscure, possibly grand, that first ignited the desire to understand. Richard Feynman spoke of fixing old radio sets. Carl Sagan describes glimpsing the stars through the city lights of New York, 'even with an early bedtime', and asking everyone he met what they were. 'They are lights in the sky, kid!' was the reply. So he went to the library on 85th Street and found a book with a 'very big thought' inside. Stars are Suns, but very far away. For me, it was the yearly reappearance of Orion in the darkening skies above Oldham. I don't know why I always enjoyed autumn. I liked the smell; wet leaves and burnt toffee. I liked the start of the football season and new school years with fresh books. And I liked lying on my back on the hills above my house with a wet nose and cold fingers and staring dizzily at red Betelgeuse. I could hear its violence; a restless star the size of a solar system signalling the onset of northern winter. Lights in the sky, kid. That's what does it for me. Such things are personal, and usually, I would imagine, unique. But very interestingly, I brushed against a memory I never had in a house in Brisbane while filming rhinoceros beetles. The owner of the house was an avid entomologist, and had assembled a small collection of butterflies, insects and beetles, neatly pinned, arranged and classified in darkly mottled glass-fronted wooden drawers in his garage. Endless forms most beautiful, diverse to the point of profligacy, natural selection as a crazed, obsessional artist, sculpting and painting for the sheer delight of creation. Celebration, classification, curiosity, consideration, comprehension, in that order, and a lifetime as a biologist follows.

Our visit to suburban Brisbane was driven by the specific, notwithstanding my glimpse of a road not taken, unexpectedly conferred upon an unsuspecting physicist by the gloriously diverse. *Xylotrupes ulysses australicus*, to give the rhinoceros beetle its full name, is a common animal in this part of Australia. During the summer months, an invasion

144

takes place, as millions of individuals with uncommon powers take up residence on every poinsiana tree and lamppost. At up to 6 cm in length, the rhinoceros beetle is difficult to miss, its size complemented by a profligate collection of spikes and horns that inspired the species' common name.

In keeping with all insect bodies, a beetle's anatomy is divided into three sections: the head, the thorax and the abdomen. All beetles have a hard exoskeleton, made up of tough but flexible plates called sclerites, while its front set of wings has been transformed into a hardened pair of shell-like structures known as elytra, which offer protection to the rear body and wings. This combination of hardened and functional wings gave rise to the name Coleoptera, from the Greek *coleos*, meaning 'sheathed wing'. Accordingly, the vast majority of beetles can fly. Despite sharing this common anatomical uniform, beetles are tremendously diverse, with an arsenal of different features and appendages that have evolved for all manner of uses. In the case of the rhinoceros beetle, it is the defensive horns that give it such a characteristic look, although only the males possess them. The nocturnal rhinoceros beetles spend much of their lives underground as larvae – a development stage measured in years rather than months. Eventually they emerge to begin an adult life that lasts for a short 4 months, during which they will breed, and it is competition for females which has driven the evolution of the male horns. They are used for grappling with other males

during the brief mating season; the females have a far less aggressive appearance, sitting patiently in the background while the males posture. The horns not only indicate the strength of the male, they are an 'honest signal' of a male's health and fitness. Rather amusingly, there are two quite different populations of male rhinoceros beetles. While the elaborately posturing alpha males are fighting it out, a smaller group lacking in physical prowess cleverly sneaks up on the waiting females, no doubt tapping their feet impatiently on a branch. I like to think that this is also the reason for the co-existence of geeks and footballers in human populations.

Despite their hostile behaviour, however, rhinoceros beetles are harmless; they rear up and hiss a lot, by contracting

........................................................................

*Gram for gram, these insects are thought to be one of the strongest animals alive... capable of supporting items 850 times their own weight.*

........................................................................

their abdomens, but this is all. Handbags at dawn, as they say on a football pitch. But before we get too contemptuous, the rhinoceros beetle does deserve respect, because, gram for gram, it is one of the strongest animals alive.

By many measures, Hossein Rezazadeh can be considered the strongest man on Earth. As one of the greatest weightlifters of all time, the Iranian-born Hossein is an Olympic champion and world record holder with a formidable record in elite competition. The highlight of his career remains the 2004 Athens Olympics, where he raised 263.5 kg above his head in the clean and jerk to win the gold medal and set a world record that, at the time of writing, still stands. This is a colossal amount of weight for a human to lift, and the fact that the record remains unbeaten eight years later suggests that this is close to the limits of human performance.

In the context of the animal kingdom, however, this seemingly superhuman strength should be put in context. Hossein's body weight when he set his world record was 152 kg. His gold medal lift was therefore significantly below twice his own body weight. Compare this Olympian performance with that of a rhinoceros beetle: a large individual weighs around 2 g, and can carry loads of over 30 times its own body weight at high speed and over long distances. That's an impressive ratio. Scaled up to human size, world record holder Hossein Rezazadeh would have to be able to lift four family cars onto his back and carry them for many kilometres.

The rhinoceros beetle's strength became apparent to me when I was persuaded to grab the horns of a large hissing male while he was sitting on a hefty broken branch and pick the whole lot up. It was heavy, but the beetle simply stiffened his legs and let the branch swing happily away below him. The horn feels like stiff plastic, by the way. ◉

**WEIGHT-TO-STRENGTH RATIO**
The weightlifter Hossein Rezazadeh would have to lift 130 tonnes to compare with the strength of the rhinoceros beetle.

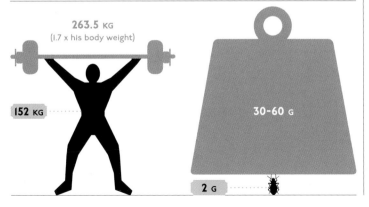

263.5 KG
(1.7 x his body weight)

152 KG

30–60 G

2 G

# BROKEN MEN AND SPLASHING HORSES

Unlike mammals, reptiles and amphibians, today's insect populations appear to have forgotten that gravity exists. Insects are the superheroes of the animal kingdom, clinging upside down to walls, or exhibiting seemingly impossible acts of strength and resilience in their ability to fall from great heights and survive. The key to understanding why these creatures appear to avoid the constraints faced by larger animals lies, yet again, in the universal laws of physics.

The history of science is replete with interesting characters. There is a definite sense of liberation, albeit in being constrained by the laws of nature rather than by the norms of polite society. Seniority and position are an irrelevance, mere opinion matters for nothing, and while there is no possibility of being absolutely correct, there is the satisfying feeling of being able to label others as certainly wrong. Whether this attracts a particular type of character, or builds it, is unclear to me, but I am certain that the above paragraph is an excellent description of myself.

When it comes to the great eccentrics in the upper end of the distribution, however, few compare to the brilliant, anarchic and downright daring J B S Haldane. Born in 1892 on bonfire night, his birthday might be seen as a metaphor for his life. Haldane was a pioneering geneticist and evolutionary biologist with a passion for self-experimentation that tested both the limits of knowledge and his own body in equal measure, but never his sense of humour. He drank copious amounts of hydrochloric acid to discern its effect on muscle. He increased the level of oxygen saturation in his blood to such a level that it initiated a fit, and he performed so many experiments in a decompression chamber that he lost count of how many times he perforated his eardrum. As with all of Haldane's experiments, the science and the danger were quickly followed by the wit, and in most cases a memorable quote would emerge alongside his results. In his famous essay 'On Being the Right Size', written in 1926 while he was teaching at Trinity College, Cambridge, he vividly described the relative importance of gravity to animals of different sizes:

*'You can drop a mouse down a thousand-yard mine shaft; and, on arriving at the bottom, it gets a slight shock and walks away, provided that the ground is fairly soft. A rat is killed, a man is broken, a horse splashes.'*
—*J B S Haldane*

Never afraid of mixing polemic with science, the great essay ends in typically tangential fashion with reflections on socialism and the possibility to implement it in organisations the size of countries. 'I find it no easier to picture a completely socialised British Empire or United States than an elephant turning somersaults or a hippopotamus jumping a hedge.'

It seems obvious, because of our experience, that small things should bounce whereas big things splash. But why is this? At first glance, one might not expect there to be any difference. All objects fall at the same rate in a gravitational field in a vacuum. The deep explanation for this is that they are following geodesics – that is 'straight lines' – in spacetime curved by the presence of the Earth. Since their paths have everything to do with the spacetime and nothing to do with their mass, it should be obvious that all objects (including, note, light itself, which has no mass at all) fall towards the ground at the same rate. This is graphically demonstrated in orbit, where astronauts and water droplets float happily together inside a spacecraft, all of which might be described as plummeting towards the Earth. In the presence of air, however, all things do not fall at the same rate. This, as Haldane notes in his essay, is due to the increasing effects of air resistance as animals

where m is the mass of the animal, g is the acceleration due to gravity, ρ is the density of air, $C_d$ is the drag co-efficient (a dimensionless quantity which depends on the shape of the animal) and A is the surface area of the animal. As the size of the animal increases, therefore, the mass increases at a faster rate than the surface area and the terminal velocity increases.

Big animals, in other words, hit the ground at a higher speed than small animals. Big animals also hit the ground with higher energy than smaller animals, because their mass is greater and their velocity is greater, and their kinetic energy is equal to the mass multiplied by the square of the velocity. But, as for the case of bones, an animal's strength is proportional to its cross-sectional area, which is greater in proportion to its mass for smaller animals. Small animals win on all fronts when falling, in other words. They hit the ground slower, the energy they must dissipate is less, and they are relatively stronger, in proportion to their mass, than large animals.

It is not gravity that defines the life of the smallest creatures; rather, it is the other long-range fundamental force; electromagnetism. We have, in fact, met the electromagnetic force before. It is the force responsible for the strength of materials – the force that holds atoms and molecules together. The fundamental reason why the strength of any given material increases in proportion to its surface area is that a larger surface area means that more molecules are 'in contact', by which we mean close enough to be strongly attracted to each other by the electromagnetic force. The challenges faced by an animal therefore represent the relative importance of gravity and the electromagnetic forces between molecules.

Watch a house fly walk up a window, or a spider scoot along a ceiling, and you may be forgiven for thinking that gravity is being defied in front of your very eyes. This would be wrong, of course. The laws of physics are never defied. Instead, you are seeing a physical manifestation of the relative strengths of gravity and electromagnetism on the animals. You can demonstrate this to yourself with the simplest of experiments. Get yourself a small piece of paper, lick your finger and hang it upside down. If the paper is not too big, it will of course stay attached to your finger. The reason is that the electromagnetic forces between the water molecules and the molecules in your finger and the paper are stronger than the gravitational force exerted by the Earth on the paper. Think of that for a moment. An entire planet is attempting to pull the paper towards it, and the forces between a few molecules can resist the pull. The deep reason for this is one of the great mysteries in physics. Gravity is an incredibly weak force, and nobody knows why.

Many insects use this effect to walk around upside down. They secrete a sticky fluid on to their feet, which allows them to adhere to even slippery windows, without falling off. No matter how much sticky fluid you attach to your hands and feet, however, you won't be able to acquire this superpower, because you are simply too massive in relation to the surface area of your hands and feet and the adhesive forces between molecules, and the weak force of gravity therefore dominates. ◉

get smaller. This is a consequence of the scaling laws we discussed earlier in the chapter. If we think in more Newtonian language, for a moment, which is appropriate, then we can picture two forces acting on a falling animal. The gravitational force, acting downwards towards the Earth, is proportional to the mass of the animal, which is proportional to the cube of its size. The drag force (air resistance) acts in the opposite direction to slow the animal down, and this is proportional to the surface area of the animal, which is proportional to the square of its size. As an animal gets smaller, therefore, air resistance becomes more important. If the animal falls for long enough, it reaches what is known as terminal velocity, which is given by

$$V_t = \sqrt{\frac{2mg}{\rho A C_d}}$$

# AS SMALL AS
# IT GETS...

Every summer when the water levels drop in Lake Clifton in western Australia, a window opens onto the world of the very small. Prokaryotes, the single-celled bacteria and archaea, dominate Earth in terms of numbers of living cells; it is estimated that there are around $5 \times 10^{30}$ individuals on the planet at any one time – that's 5 million million millon millon million cells, containing as much carbon as Earth's entire population of plants. There are a billion bacterial cells in a litre of drinking water, and of course they are absolutely invisible to the naked eye. Yet here on Lake Clifton the dead cells of the smallest lifeforms lead to the appearance of great structures, yellowing mound-like formations that break the water surface along the shoreline. They look almost geological, although there is a hint of the living about them, and indeed they are created through the growth of colonies of bacteria; life moulding the land. They are known as thrombolites, a name that derives from the clotted structure created by calcium-rich bacteria as they grew on each other in the lake over hundreds of years. Thrombolites flourished in the Cambrian and lower Ordovician (540–470 million years ago), when calcium became widely available in the sea. They were preceded, billions of years earlier, by stromatolites, bulbous mats of bacteria that trapped sediment and produced similar-looking lumps of rock. Although rare today – perhaps because those microbial mats tend to get eaten by animal grazers – stromatolites would have been a common sight in Earth's seas for billions of years. For much of the history of our planet, the only macroscopic sign of life would have been structures like these rocks.

It is easy to dismiss bacteria as simple organisms; basic compared to the complexity of the eukaryotes. Yet these organisms are far from simple. Each individual houses all the necessary complexity of life. DNA, RNA, mRNA, ribosomes, all the basic biochemical machinery fundamental to every living thing on the planet, whirs away within their cellular membranes. This raises an interesting question. What is the smallest possible living thing? The smallest bacteria are members of the genus Mycoplasma, measuring around 0.2 microns, two ten thousandths of a millimetre. This is comparable in size to the largest viruses, which are not considered to be alive by most biologists because they cannot reproduce without the aid of their hosts' cellular machinery. On Earth, at least, 0.2 microns seems to be the minimum size required for a free-living, self-replicating machine. Is this the result of the peculiarities of the evolution of life on Earth, or is there a deeper reason for this limiting size?

The size of a machine built of molecules ultimately depends on the size of the molecules themselves, and this in turn depends on the size of atoms. A carbon atom has a radius of approximately 0.2 nanometres, which is a thousand times smaller than the Mycoplasma. In other words, you could line around 1,000 carbon atoms along the smallest bacterium cell. In 2006, the symbiotic bacterium *Carsonella ruddii* was found to contain only 182 different proteins. Current estimates suggest that at least 100 different kinds of protein molecule are required to construct a living cell, along with the associated genes, RNA and molecular machinery necessary to control their production. All this has to fit into a sphere of radius 0.2 microns. Some life on Earth, therefore,

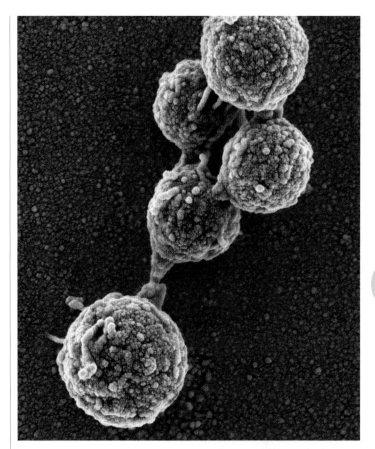

*The strange formations that stud this shoreline are created through the growth of colonies of bacteria, a beautiful example of life moulding the land.*

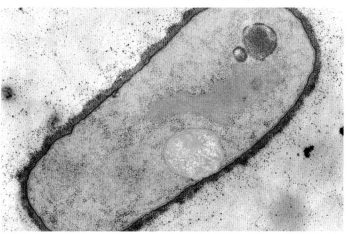

probably exists at the minimum size limit, which is ultimately determined by the size of atoms and molecules, and therefore by the laws of physics. At the most fundamental level, the size of an atom is determined by a handful of physical constants such as the strength of the electromagnetic force, the mass of the electron, and Planck's constant. If some of the more esoteric physical theories are correct, there may be many other universes with different values of these constants. In many of these universes, atoms could not exist, and therefore life would be impossible. But in our Universe, all atoms, everywhere, are the same size because the fundamental constants of nature are the same, everywhere. This imposes a non-negotiable limit on the size of living things in our Universe, wherever they may be found, and that limit will be close to 0.2 microns, the size of the smallest free-living bacterium cells on Earth. ◉

**LEFT:** Thrombolites at Lake Clifton, western Australia, are formed from the dead cells of the smallest life forms.

**TOP:** *Mycoplasma mycoides*, one of the smallest known cells.

**ABOVE:** *Bacillus megaterium*, a rod-shaped bacterium.

# THE SMALLEST MULTICELLULAR LIFE ON EARTH

**BACTERIA**
*0.5–5.0 micrometers*
$(10^{-6})$

At a magnification of ×64,000 size, this coloured transmission electron micrograph shows T2 bacteriophage viruses (in orange) attacking an E. coli bacterium.

**MYCOPLASMA**
*0.1 micrometers*

At a magnification of ×71,000 size, this coloured transmission electron micrograph shows a section through cells of *Mycoplasma pneumonia*.

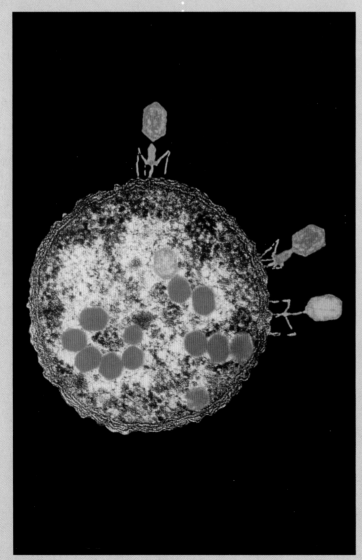

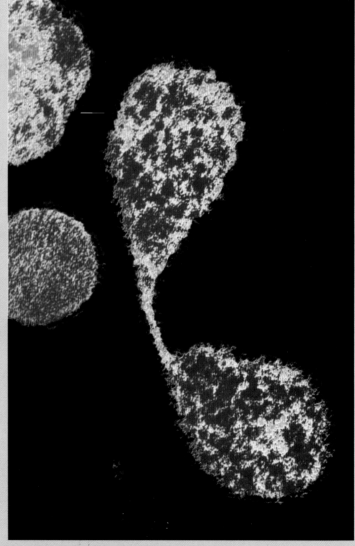

## PROTEIN
*0.004 micrometers*

Computer graphic
of a molecule of the
protein somatotropin,
the human growth
hormone, produced
by the anterior lobe
of the pituitary gland.
It is a large molecule,
being formed of 191
amino acids and 22 kDa
(kiloDaltons).

## DNA MOLECULE
*0.0025 micrometers*

In living organisms
DNA usually exists as a
pair of molecules that
are held tightly together
– these two long strands
entwine in the shape of
a double helix.

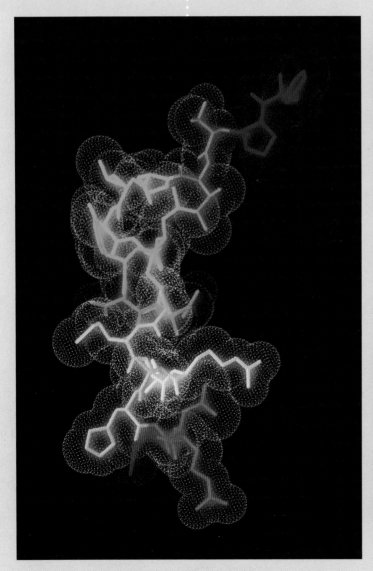

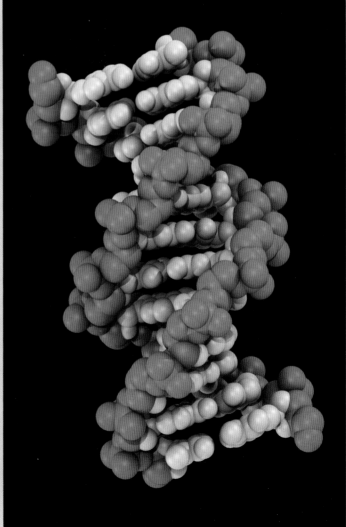

# SIZE REALLY MATTERS

**BELOW:** Descending into the Naracoorte Caves, South Australia. The caves are home to the endangered little bent-wing bat.

Walking in the Naracoorte Caves in South Australia is a precarious business. The cave system was created over millions of years by running water, forming a Byzantine system infamous for its potholes and hidden crevices. In the distant past some of the continent's long-lost creatures stumbled into these caves, where they were trapped and preserved, making Naracoorte one of the richest fossil sites for large fauna in Australia.

We came to Naracoorte to film one of Australia's most endangered animals, the little bent-wing bat (*Miniopterus australis*). There are only two known caves in which these bats rear their young. Inside, it is estimated that there are approximately 35,000 individuals, of which a third are young bats. We have to be exceptionally careful not to disturb the colony, not least because these tiny members of the Miniopterus genus breed just once every summer, with a single pup.

The little bent-wing bats certainly live up to their common name; when fully grown, they are just 4 cm long, and weigh little more than 20 g. Roosting deep in the cave system during the day, they emerge at dusk in their thousands to feed. In a single night the bats travel up to 80 km from the cave in search of food, returning at dawn to nourish their young.

Bats, like all mammals, are endotherms – they are warm-blooded creatures that generate heat internally to maintain their body temperature. It's an evolutionary adaptation they share with thousands of other species, and with it comes enormous freedom. Unlike their cold-blooded counterparts, warm-blooded animals are not necessarily tied to the changing temperatures of the day or the shifting influence of the seasons, although little bent-wing bats hibernate during the winter months when insects are scarce. It is the ability to control the internal temperature in a wide range of external conditions that enables mammals and other endotherms to

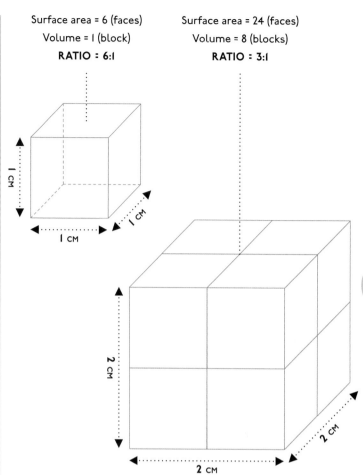

Surface area = 6 (faces)
Volume = 1 (block)
**RATIO : 6:1**

Surface area = 24 (faces)
Volume = 8 (blocks)
**RATIO : 3:1**

**BELOW:** The little bent-wing bat needs to eat a lot of insects – often as much as its own body weight every night – simply to stay warm.

populate so many different environments on the planet, and to live with such independence. Warm blood, however, comes at a cost. Maintaining body temperature takes enormous effort, because the heat energy has to come from somewhere. Many warm-blooded animals have a larger number of mitochondria per cell than their cold-blooded counterparts. This allows them to increase the rate at which their bodies burn their food reserves in oxygen, and so generate more energy, as long as they have sufficient access to food. A little bent-wing bat may eat in excess of its own body weight in insects each night, simply in order to stay warm.

The reason for this extreme demand for food, which drives much of the bat's behaviour, is its small size. All endotherms lose a large fraction of the heat they generate through the surface of their bodies. This is what makes a packed room warm up so quickly; a resting human adult is effectively a 100-watt heater. The dominant factor that determines the rate of heat loss at a given temperature is the surface area to volume ratio of the animal. This is easy to see; the total amount of heat produced by an animal will be determined roughly by the number of cells in its body, and this is proportional to its volume. Heat is lost through the surface of the body, by radiation and conduction. As we have already seen, an animal's volume is proportional to the cube of its size, whereas its surface area is proportional to the square of its size. As animals get bigger, therefore, their volume increases proportionally faster than their surface area, and they find it much easier to maintain a high internal body temperature.

In fact, for large animals, the problem is not heat loss so much as heat build-up; large endotherms are in danger of cooking themselves from within because they cannot dissipate heat fast enough from their surfaces, and this is one of the limiting factors that constrains the size of the largest animals. For animals as small as the little bent-wing bat, however, the rate of heat loss is bordering on the prohibitive, because their surface area is proportionally so large in relation to their volume.

Hidden away inside their cave, the bats engage in complex social behaviours just to retain heat and stay alive. They huddle together on the ceiling in large groups, effectively decreasing the surface area to volume ratio of the colony as a whole, and therefore retaining more of the heat they produce. Think of the huddle, if you like, as a single larger organism with the volume of many hundreds of bats.

Animals are not completely powerless in the face of the relentless mathematics of the surface area to volume scaling law, however. They are able to vary the amount of energy they produce by varying their metabolic rate. Bats, like all small warm-blooded animals, have a high metabolic rate in order to maintain their body temperature. They breathe rapidly, their hearts beat fast and as a consequence they must eat a tremendous amount – a life lived at full speed in every sense. Larger endotherms, such as human beings, have a lower metabolic rate. The relationship between the mass (and therefore volume) of an animal and its metabolic rate, however, is not what might be expected from the surface area to volume scaling law.

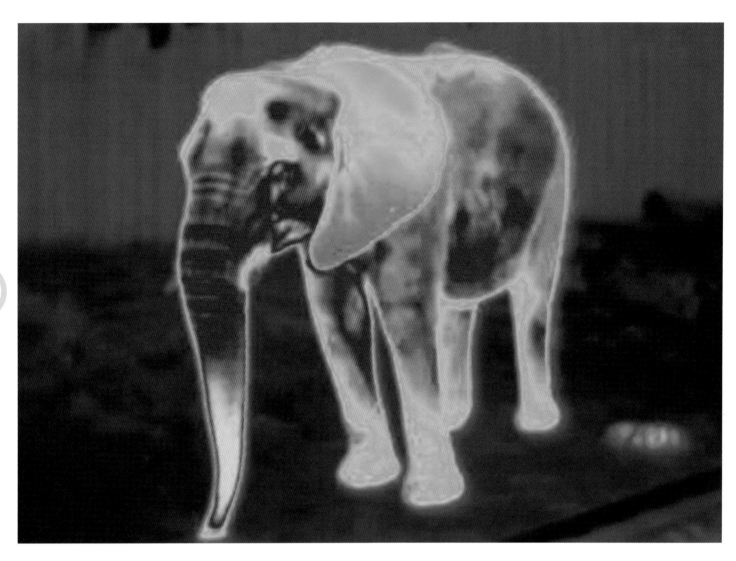

Take a look at the illustration opposite; it is a graph of the rate of energy use (in watts) for a wide range of mammals plotted against their mass. If the amount of energy an organism produces through its metabolic processes increased in direct proportion to its mass, we would expect to see a straight line on the graph with a slope of 1. We can do better than that, however, because we know about the surface area to volume scaling law; we might hazard a more informed guess, therefore, that the slope should be something like 2/3. This is not what is observed. There is clearly a scaling law in operation, but the slope is very close to 3/4.

Calculating the precise value of this number has caused a great deal of controversy in biology over the years. What is uncontroversial, however, is that the slope is less than one. This means that, broadly speaking, large animals have slower metabolic rates than smaller animals. The question is: why? There are two possible answers. One is that there is a constraint at work, which limits the metabolic rate of each cell in a large animal. This could be due to the way that supply networks such as blood vessels branch out from the animal's core to its extremities. Just like a tree, the networks carrying oxygen and nutrients around an animal's body continually branch into smaller and smaller supply vessels, and as more and more branching occurs, the supply of 'fuel' to the cells at the edge of the network becomes compromised. The cells in

| LIFETIME HEARTBEATS & ANIMAL SIZE | | | | | |
|---|---|---|---|---|---|
| Creature | Weight (grams) | Heart rate (/minute) | Longevity (years) | Product | Lifetime Heartbeats (billions) |
| Human | 90,000 | 60 | 70 | 4,200 | 2.21 |
| Cat | 2,000 | 150 | 15 | 2,250 | 1.18 |
| Small dog | 2,000 | 100 | 10 | 1,000 | 0.53 |
| Medium dog | 5,000 | 90 | 15 | 1,350 | 0.71 |
| Large dog | 8,000 | 75 | 17 | 1,275 | 0.67 |
| Hamster | 60 | 450 | 3 | 1,350 | 0.71 |
| Chicken | 1,500 | 275 | 15 | 4,125 | 2.17 |
| Monkey | 5,000 | 190 | 15 | 2,850 | 1.50 |
| Horse | 1,200,000 | 44 | 40 | 1,760 | 0.93 |
| Cow | 800,000 | 65 | 22 | 1,430 | 0.75 |
| Pig | 150,000 | 70 | 25 | 1,750 | 0.92 |
| Rabbit | 1,000 | 205 | 9 | 1,845 | 0.97 |
| Elephant | 5,000,000 | 30 | 70 | 2,100 | 1.1 |
| Giraffe | 900,000 | 65 | 20 | 1,300 | 0.68 |
| Large Whale | 120,000,000 | 20 | 80 | 1,600 | 0.84 |

**LEFT:** Thermogram of an elephant, showing the coldest areas in blue, and the warmest in red.

**RIGHT:** In comparison to the elephant (opposite), this thermogram of a mouse shows warm areas in yellow, indicating that it has a higher metabolic rate than the elephant.

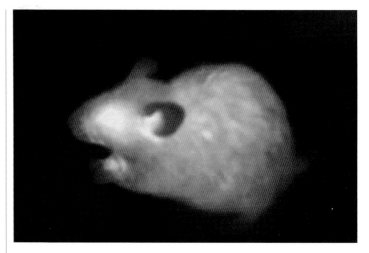

## ENERGY USE IN MAMMALS

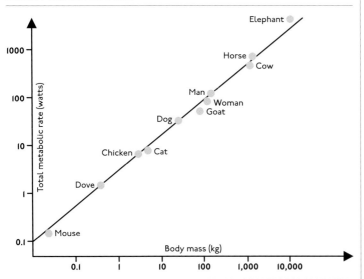

## RELATION BETWEEN REST HEART RATE AND LIFE EXPECTANCY IN MAMMALS

*Just like a tree, the networks carrying oxygen and nutrients around an animal's body continually branch into smaller and smaller supply vessels.*

large animals, with highly branched networks, may therefore have been forced to operate at a lower metabolic rate. It has been shown mathematically that networks branching in this way give rise to quarter-power scaling laws such as those observed in the metabolic rates of animals.

The other answer is that the surface area to volume scaling law presented an opportunity for large animals. Because they retain more of their internal heat, their cells evolved to run at a lower rate because they could. They took advantage of an opportunity, in other words, which resulted in them having to eat less and therefore spend more of their time on other beneficial activities such as mating or rearing their young.

The benefits of a lower metabolic rate stretch beyond lifestyle, however. In mammals, the number of heartbeats in a lifetime appears roughly independent of size or species. But mammals with higher metabolic rates have higher heart rates, and indeed are observed to have shorter lifetimes. Notice that we don't draw the inference that the heart itself is somehow limited to a particular number of beats. This simple invariant number is probably a reflection of deeper molecular processes, perhaps relating to the rate of synthesis of ATP in each cell. But whatever the reason, there is an interesting relationship which seems to demand an explanation.

There may be simpler reasons for the observed relationship between body mass and metabolic rate. Large animals have bigger bones, as we have seen, and so it could simply be that large animals have a greater proportion of their mass taken up by inert structures that use no energy. The observed metabolic rate would be lower, therefore, than might be expected from a simple scaling law that assumes a one-to-one correspondence between the number of active cells in an animal and its volume (and therefore mass).

The empirical data, however, are interesting and of more than passing interest to large animals such as ourselves, because ultimately, the bigger you are, the longer you live. ◉

# AN ISLAND
# OF GIANTS

We ended our journey through the sizes of life in a quite spectacular and isolated place, with frankly the most terrifying airstrip I have ever experienced, perched cavalierly on a hilltop lashed by the unobstructed Indian Ocean winds. This ridiculous position, and ridiculous is the word I used as we approached it at an interesting angle of attack, is in fact the only place it would fit; the island is barely 16 km from tip to tip. Named Christmas Island, when it was spotted on Christmas Day 1643, this barren subtropical rock is an outpost of Australia closer to Indonesia than to the Australian coast. Its isolation has resulted in a unique ecosystem dominated by crabs, and the phenomenon of island gigantism has delivered one of the wonders of the natural world: the quite magnificent Christmas Island robber crab (*Birgus latro*), the largest land crab anywhere on the planet.

**ABOVE LEFT:** Christmas Island in the Indian Ocean plays host to the red crab (*Gecarcoidea natalis*), a species that is endemic to the island and which is seen here in vast numbers, coming ashore and climbing the cliffs beyond the beach.

**ABOVE:** The phenomenon of island gigantism is demonstrated by the enormous robber crab, the largest terrestrial arthropod in the world.

*The Christmas Island robber crab is the largest land crab anywhere on the planet and is so supremely adapted to life on land that it can even climb trees.*

Growing to over 50 cm in length and weighing over 4 kg, these animals are the largest native inhabitants of the island. According to legend, they have been known to attack the islanders' goats in the dead of night, but they live primarily on a diet of fruit and coconuts. They are extremely intelligent, at least for crabs, and have adapted well to human interaction; indeed, they seem to enjoy it. They are known locally as robber crabs, because they have a reputation for curiosity and for stealing things. They wander into unlocked houses and steal knives, forks and even shoes. I encountered this magpie-like behaviour first hand when one of the crabs opened my camera bag and stole a 5-dollar note.

The robber crabs inhabit many different worlds throughout their lives. Despite appearances, they are a species of hermit crab. Independent life for a robber crab begins as a small larva, one of thousands which are released into the sea by the female, who carries her eggs down to the shore as they are about to hatch. The larvae grow for a period of around a month, swept around in the ocean currents, until a very few are washed back into the shallows around Christmas Island where they find a shell and eventually make their way back onto the land, losing the ability to breathe in water. After around five years, having outgrown their borrowed shells and developed their hardened abdomen, the robber crabs reach sexual maturity. They then roam the tiny island for up to 70 years, making them both the largest and longest-lived crabs on Earth. ◉

**BLUE CRAB**
*Discoplax hirtipes*

SIZE
Carapace up to 16 cm
wide, at a weight of up
to 500 g

LIFE EXPECTANCY
Up to 5 years

**ROBBER CRAB
OR COCONUT CRAB**
*Birgus latro*

SIZE
Body length up to
50 cm, at a weight of
up to 4 kg, and a leg span
of more than 0.91 m

LIFE EXPECTANCY
Up to 60 years

**RED CRAB**
*Gecarcoidea natalis*

SIZE
Carapace up to
12 cm wide

LIFE EXPECTANCY
Up to 30 years

160

*The vast range in the size of life on Earth is a result of evolution by natural selection constrained by the laws of nature. There is a minimum size, which is set ultimately by the size of atoms and molecules, and there is a maximum size, which on land is set by the size and mass of our planet, because the force of gravity restricts the emergence of giants.*

*But within that framework, evolution has conspired to produce a huge array of animals and plants, each beautifully adapted to exploit the niches available to them. Your size influences your form and construction, it determines how you experience the world, and ultimately, be you shark, bat, human or Christmas Island robber crab, it determines how long you have to enjoy it.*

161

# CHAPTER 4

---

# EXPANDING UNIVERSE

# THE EXPANDING UNIVERSE

The first life appeared in a meaningless Universe filled with information. Sun and Moon illuminated young landscapes swept with un-beautiful forms, cloudless nights were punctuated with patternless stars, chemicals mingled unnoticed with atmospheric gases and ocean waves roared on sterile shores newly carved from razor-sharp rocks that posed no danger. The first living things were isolated; unable to respond to, let alone perceive, the wider world. In this chapter we will explore how living things learned to touch the rocks, smell the air and see the stars, and in doing so evolved the ability to perceive their Universe and fill it with meaning.

# PLUGGING IN

For such a small island, Santa Catalina has a rich history. Since its original inhabitants arrived almost ten thousand years ago, this quirky place, with a city called Avalon, has been home to an endless assortment of explorers, from the indigenous Californians – the Tongva – to Portuguese pioneers, Chinese smugglers, Russian fishermen and Alaskan seal hunters. They came to harvest the cool Pacific waters, chilled by the nutrient-rich California current that rises from the depths to support subtropical densities of marine life. Diving off Catalina was an experience I will not forget. From beneath the waves, the kelp forests, bright orange Garibaldi fish and swirling California sea lions feel out of place just an hour from the great industrial docks of Long Beach.

We came here to film a reclusive creature with a wonderfully inaccurate name: the mantis shrimp (*Odontodactylus scyllarus*). This creature is neither shrimp nor mantis, although it resembles both. It is in fact a separate member of the crustacean family, and more correctly termed a stomatopod. Stomatopods are abundant across the Pacific and Indian oceans and are a common ingredient in seafood dishes from Japan to the Mediterranean. Yet despite their gastronomic popularity, these creatures are difficult to film in the wild.

**BELOW:** Diving off Santa Catalina island among the giant kelp forests (*Macrocystis pyrifera*) is an unforgettable experience, exemplifying the rich and diverse marine life off the coast of southern California.

Hidden away in sandy burrows on the ocean floor, stomatopods are certainly enigmatic, rarely venturing out of the protection of their homes, spending much of their time in monogamous relationships, and living for decades. This might seem an unusual lifestyle for something that resembles a giant prawn. Stomatopods are predators, and are equipped with one of nature's most powerful weapons – one that demands respect, even from creatures as large as divers. In Australia they are known as thumb splitters.

Filming the mantis shrimp involves sitting in the muddy shallows 15 m below the surface of the Pacific and waiting in quite un-Californian cold. Eventually, encouraged by a smattering of shellfish bait, an elaborate riot of yellow mouthparts, bright blue legs and precariously balanced eyes emerges, and scurries off along the silty floor. It is a surprisingly large animal – about the length of my forearm

– and I am well aware that I should stay out of its way as I try to snare it in a net and bring it to the surface for filming. The reason for my caution is its heavily modified front legs, which it deploys like medieval clubs. The 'smasher' family of mantis shrimps uses these legs to defend territory and to crack open an assortment of prey, from rock oysters to crabs and sea snails. The key to the effectiveness of these innocuous-looking appendages is the speed of deployment. By winding up their muscles like a coiled spring, they are able to unfold their forelegs with 'astonishing' speed, throwing them forward in a blinding flash. I don't like the word astonishing as a rule, but in this case it is appropriate, because I have never seen anything quite like the ultra-slow-motion footage of a mantis-shrimp punch. It is the fastest-known movement in the animal world – as the legs unfold they reach speeds of over 80 km/h in virtually no time at all, with an acceleration of 10,000 $g$

– roughly that experienced by a bullet. This extreme acceleration creates a shock wave so powerful that the water temperature along the shock front rises to several thousand degrees Celsius. Even if the initial punch is misplaced, the shock wave alone can stun or even kill the prey, or shatter a glass tank.

The objects of our attention, however, are the mantis shrimp's eyes, which have evolved partly to allow the animal to deploy these formidable weapons with precision. We humans have binocular vision, viewing the world from slightly different angles as a result of the wide spacing of our eyes, and this allows us to estimate the distance to an object. The mantis shrimp has compound eyes, each made up of thousands of individual lenses, all pointing in slightly different directions. This allows each eye to form three images from slightly different perspectives, giving it trinocular vision and extremely accurate depth perception.

The mantis shrimp also has an extraordinarily intricate colour vision system. Humans have three-colour vision, delivered by three types of photoreceptor cells called cones, while rods – a fourth type of cell – deliver low-light sensitivity in black and white. The mantis shrimp's eyes trump this system easily, with 12 colour receptors, each of which is sensitive to slightly different wavelengths of light. The reason why these

*It is such an effective mechanism of attack that, even if the initial punch is misplaced, the resulting shock wave can be enough to stun or even kill the prey; it can even shatter the glass of a tank.*

**OPPOSITE:** Saying hello to a mantis shrimp – from a safe distance. Its front legs are incredibly powerful: small wonder that in Australia they are known as 'thumb splitters'.

**ABOVE:** The compound eyes of a mantis shrimp have evolved partly so that they can use their formidable forelegs with astonishing speed and great precision, to stun or kill their prey.

animals need such precision colour vision is not fully understood, but the peak wavelength sensitivities of the photoreceptors appear to correspond to the different colours on the mantis shrimp's body. These colourful displays are used for communication, and it may be that 12-colour vision allows for the precise identification of signals in variable water and lighting conditions; avoiding misunderstandings is important when you have the fastest punch in the natural world. Such reasoning is instructive, because it provides a hint of things to come. Each animal has evolved its particular suite of senses for a reason; they are adapted to its environment and way of life. In the case of the mantis shrimp, range-finding is an important component of its hunting behaviour, and high-precision colour vision is probably important for communication between individuals and the identification of prey in an environment commonly filled with silt and shifting illumination.

The evolution of sensory organs as complex as the mantis shrimp's eyes, fine-tuned with such precision to its particular needs, might seem difficult to explain, and indeed the evolution of the eye has achieved almost totemic status among those who hope to avoid a scientific explanation for the wonder of life. This chapter will show precisely why such concerns are misplaced. The development of the senses is instead a textbook example of evolution in action. At a biochemical level, there are startling similarities between our eyes and the eyes of the mantis shrimp, betraying our common evolutionary past. Similarly, the details of our hearing are intimately related to the physiology of our distant, water-based ancestors. This points to a deep truth in evolutionary biology. The form and function of organisms can be fully understood only when viewed in the context of their past. They are, a physicist would say, four-dimensional things; the three-dimensional structure of an animal and its senses today represents a single slice through a four-dimensional object that extends deep into the past. Trying to understand how and why our eyes or ears function the way they do without reference to our deep past is like trying to understand how a mobile phone works by running it through a salami slicer and referring to only a single micron-thin sliver. But seen in all their glory, with modern, magnificently adapted forms set alongside their evolutionary origins, the story of the senses becomes more fascinating, more understandable and more wonderful than the somewhat baffling picture delivered by a consideration of only the micron-thin sliver of the present. ◉

**BELOW:** Mantis shrimp have such powerful front legs that they have been known to shatter the glass of a tank.

**RIGHT:** The compound eye of a mantis shrimp has twelve colour receptors, which allow for the precise identification of signals in variable water and light conditions.

## SPECTRAL SENSITIVITY

### STOMATOPODS

Twelve colour receptors which are sensitive to different wavelengths of light

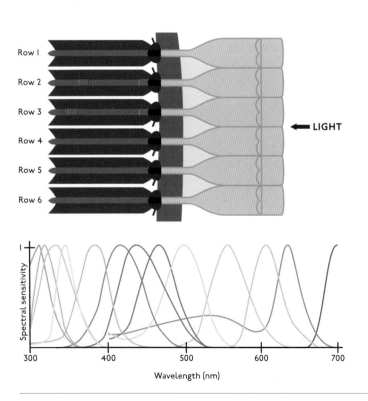

Row 1
Row 2
Row 3
Row 4
Row 5
Row 6

← LIGHT

Spectral sensitivity

300   400   500   600   700
Wavelength (nm)

### HUMANS

Three colour receptor cells – cones – alongside rods, which deliver low-light sensitivity in black and white

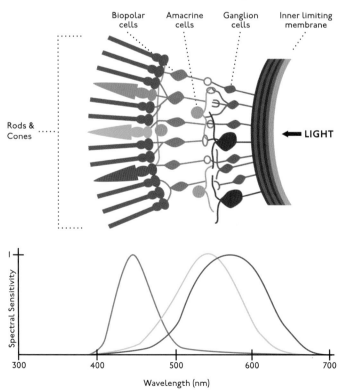

Biopolar cells    Amacrine cells    Ganglion cells    Inner limiting membrane

Rods & Cones

← LIGHT

Spectral Sensitivity

300   400   500   600   700
Wavelength (nm)

# THE COMMON SENSE

The first living things that we would recognise as complex – in the sense that they were multicellular organisms with recognisable body plans – probably arose around 650 million years ago. Before these earliest Ediacaran fossils, there is no evidence of multicellular life on Earth. By the time of the Cambrian explosion 530 million years ago, there were undoubtedly animals that could sense the world and respond to it in much the same way that we do. But the basic biochemical processes that underpin all animal senses certainly predate the emergence of multicellular life. A powerful way to investigate the evolutionary history of a particular mechanism or trait is to survey the living things today that are known to share that mechanism, and then to search for an earliest common ancestor. In the quest for the origins of the senses, we don't have to look very far, because lurking within the most innocuous pools and puddles of water across the planet is a creature that gives us a beautiful insight into how life first reached out to touch the Universe.

## THE PARAMECIUM

Take a look at the picture below. It is a paramecium, one of a large group of organisms called protists. Just like the cells in our own bodies, the paramecium is a eurkaryotic cell, with a nucleus (in fact most paramecia have two) containing the cell's genetic material, and a host of intracellular machinery contained within the cytoplasm. In common with most protists, paramecia are single-celled organisms. We share a very distant common ancestor with these unicellular life forms. We may have been on separate evolutionary paths for around 1.4 billion years, yet the basis of these creatures' interaction with the world is similar to our own. Watch a sample of these organisms under a microscope and it is clear that these tiny cells – just 100 microns across – are able to sense and respond to their environment. The outer surface of each paramecium is covered in a mass of tiny hairs called cilia, which allow it to swim around its watery habitat. These hairs are embedded in the membrane of the cell wall and beat frantically, moving the cell around on an apparently random path in the search for food. This is no haphazard trajectory, however; as soon as the paramecium bumps into an object, the cilia reverse, changing the direction of the organism. This is a simple response; the paramecium has no nervous system or brain, and yet it has a rudimentary sense of touch; when it bumps into something, it changes its behaviour. The biochemical mechanism underlying the paramecium's touch response is known as an action potential. It is near-universal, appearing in animals, protists and even some plants – implying that it is very ancient. It is also, quite fascinatingly, electrical. ◉

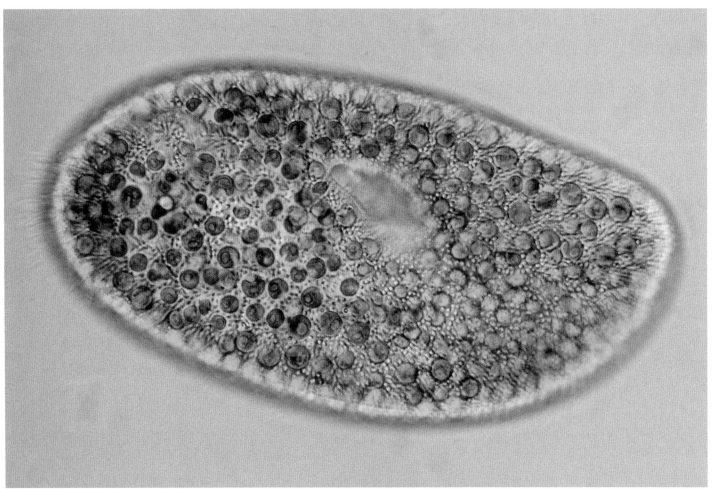

**BELOW LEFT:** Like most protists, paramecia – such as the one shown here – are single-celled organisms, yet humans share a very distant common ancestor with them.

**BELOW:** This paramecium is covered in tiny hairs, called cilia, which allow it to swim, and which also provide it with a rudimentary sense of touch.

# A BOLT FROM THE BLUE

For over 80 years, a single image has adorned the face of the United States $100 bill. Sometimes referred to as a 'Benjamin', the largest US banknote honours an individual who is widely recognised as one of the founding fathers of the United States. Until his death in 1790, Benjamin Franklin lived a life of extraordinary political and scientific achievement. It seems there were few things he wasn't good at; politician, author, newspaper proprietor, musician, statesman, scientist and inventor are just a handful of the entries on his CV.

Franklin, like many inquisitive minds of the mid-eighteenth century, was intrigued by electricity. It had been suggested, but not proven, that electricity was responsible for one of the more spectacular and destructive natural phenomena – lightning. At the time, the biggest sparks that could be created under laboratory conditions were only a few centimetres long, but Franklin was intrigued to see whether the streaks of lightning he saw across the sky were also made

*Franklin designed an experiment that would allow 'the electrical fire, to be drawn out of a cloud silently...'*

of the 'electrical fluid', as it was known at the time. To prove his hypothesis, Franklin designed an experiment that would allow 'the electrical fire, to be drawn out of a cloud silently...'.

Experimenting with lightning is a dangerous game at the best of times, but in the mid-eighteenth century the risks were significantly enhanced by a complete ignorance of the underlying phenomenon. In 1750 Franklin published a description of an experiment that involved flying a kite into a thunderstorm. In the worst traditions of modern British broadcasting, one cannot resist advising that this should not be tried at home (even though this book is fundamentally Darwinian). Although there is no substantiated record of Franklin actually performing his experiment, it is thought that he may have conducted the kite test in Philadelphia in June 1752, and successfully extracted electrical sparks from the gathering clouds. What is certain is that many others attempted Franklin's experiment, including his friend and colleague Thomas François Dalibard, who used a 12 m iron rod instead of a kite and survived. Professor Georg Wilhelm Richmann was less fortunate. He became the first scientist to experience the true power of lightning, albeit briefly, when he was electrocuted while performing Franklin's experiment in St Petersburg just a few months later.

The precise mechanisms by which thunderstorms form are still an active area of research today, but for our purposes they demonstrate a simple physical process that is central to the operation of action potentials: charge separation. Thunderstorms form when moist air rises, transporting water vapour into the cooler upper atmosphere. As the rising air cools, water vapour condenses out to form water droplets.

**LEFT:** Cloud-to-ground and aerial lightning above Tuscon, Arizona. How thunderstorms form is still an active area of research, more than 260 years after Benjamin Franklin first began experimenting with lightning.

**BELOW:** Benjamin Franklin and his son, William, fly a kite in a thunderstorm, in order to 'draw the electrical fire' out of the clouds.

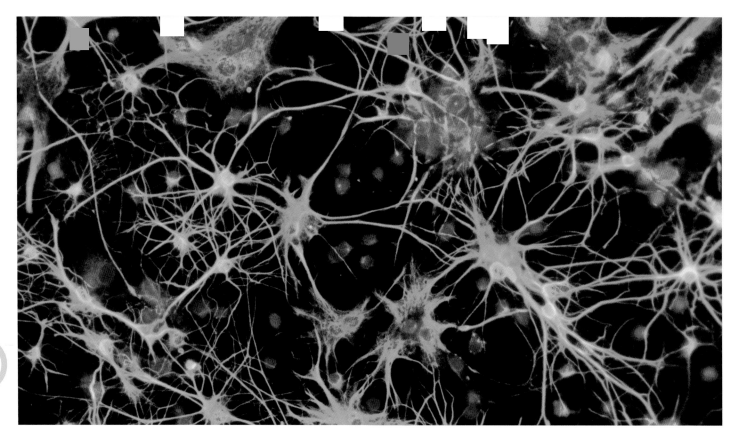

If this unstable air continues to rise and the temperature drops below freezing, ice crystals will also form. The water droplets and ice crystals acquire electric charge through frequent collisions, and the positively charged icy particles tend to be driven towards the top of the cloud while the negatively charged particles sink. This is charge separation; there is an imbalance of electrical charge between the top and bottom of the cloud, and between the bottom of the cloud and the ground, known as a potential difference. As we have already seen in Chapter 2, nature abhors a gradient, and any imbalance will tend to be corrected. In the case of a thundercloud, this correction occurs in the form of a bolt of lightning – a spark, equalising the potential difference between clouds, within clouds, or, most dangerously, between the cloud and the ground. We have been deliberately vague in our description of the process of charge separation inside thunderclouds because it is extremely complex and not fully understood. But the important point is that charge separation does occur and this creates a potential difference – otherwise known as a voltage difference – which nature will always equalise if it can.

### THE PULSE-SHAPING ABILITY OF THE PARAMECIUM: ACTION POTENTIALS

Virtually all eukaryotic cells actively maintain a potential difference across their outer membranes. This is known as a membrane potential. They do this by pumping electrically charged ions across the cell membrane. The classic example is the enzyme known as $Na^+K^+$-ATPase, which actively pumps sodium ions out of the cell and potassium ions into the cell. Both ions are positively charged, but potassium ions are allowed to leak back out of the cell again, against

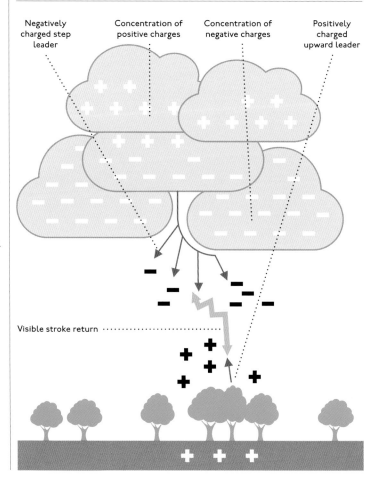

**A TURBULENT DYNAMIC:** A typical lightning strike can bridge a potential difference of several hundred million volts.

Negatively charged step leader

Concentration of positive charges

Concentration of negative charges

Positively charged upward leader

Visible stroke return

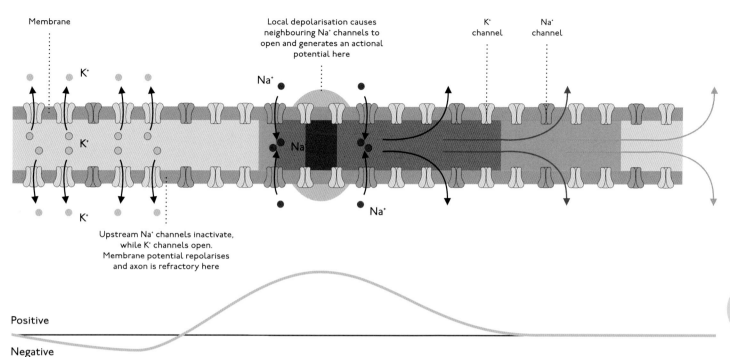

Membrane

Local depolarisation causes
neighbouring Na⁺ channels to
open and generates an actional
potential here

K⁺
channel

Na⁺
channel

Na⁺

Na

Na⁺

K⁺

K⁺

K⁺

Upstream Na⁺ channels inactivate,
while K⁺ channels open.
Membrane potential repolarises
and axon is refractory here

Positive

Negative

the concentration gradient, through potassium-selective leak channels. This establishes a potential difference between the inside and outside of the cell; the interior of the membrane ends up with a net negative charge, of just under a tenth of a volt. There are many other ions involved in the setting up and maintenance of the membrane potential, and the details are slightly different among animals, plants and other eukaryotes such as the paramecia. But the principle of maintaining a potential difference across the cell membrane by active charge separation is ubiquitous, and certainly a very ancient development. Cells use their membrane potential for two main purposes; one is simply as a battery, powering different processes in the membrane itself. The other is for the transmission of signals via action potentials, which lie at the heart of all animal senses.

Positively charged sodium ions (Na⁺) are pumped out of the cell, and positively charged potassium ions (K⁺) are pumped into the cell. The potassium ions are allowed to leak back out again through potassium-selective leak channels. This allows a net negative charge to build up along the interior side of the membrane, and a positive charge to accumulate on the outer side.

When a paramecium bumps into an obstacle, its cell membrane is deformed. This opens ion channels, similar to the potassium-selective leak channels. For the specific case of the paramecium, deformation of the cell membrane opens up voltage-dependent calcium channels, which allow calcium ions to flood back into the cell. This reduces the potential difference across the membrane, causing the membrane potential to become less negative. The channels are known as voltage dependent because they close as the membrane potential swings above zero and becomes positive. As the

calcium channels close, voltage-dependent potassium channels open, allowing potassium ions to leave the cell, re-establishing the membrane potential. This process is shown in the diagram above, and the result of all this complexity is a carefully shaped pulse, which lasts for several milliseconds. It is this pulse that is known as an action potential. In the paramecium, the action potential is used very directly to control the beating of the cilia. At lower calcium concentrations, the cilia beat forwards, but as the calcium rises, they reverse, and the paramecium changes direction. The length of time the cilia beat in the opposite direction is controlled by the duration of the action potential, which is finely tuned by the different voltage thresholds and the arrangement of ion channels as they open and close. The reason for this rather daunting complication is that a rather complicated task has been accomplished! A very precisely shaped electrical pulse has been produced in response to an external stimulus of variable length and intensity – in this case a collision between the paramecium and an external object. Such precise pulse shaping is a marvel of analogue electronics, something that we learnt to do in electrical devices only a few decades ago, and all accomplished by the selective movement of charged ions across a cell membrane. ◉

**TOP LEFT:** Immunofluorescent light micrograph of brain cells from the cortex of a mammalian brain. The nucleus of each cell is stained blue and the cytoplasm is stained green. The star-shaped cells, known as astrocytes, have numerous branches of connective tissue which provide support and nutrition to the nerve cells.

We don't normally think of plants as having senses. But they do respond to the environment around them – they grow towards the Sun, they send roots down into the Earth, and some even show a dramatic sense of touch.

This image shows a mimosa. Mimosa is a sensitive plant; when something brushes past it, it is able to detect the contact and respond by rapidly folding up its leaflets. This response was first studied by the great seventeenth-century English scientist Robert Hooke, who was interested in whether plants might have nerves. Now, plants do not have nerves. But the mimosa's response is still a sense of touch, and it is triggered in the same way as ours and that of the paramecium. If we plug electrodes into the plant around the base of the leaves, we can even measure the electrical surges.

When I stimulate the leaves repeatedly, a little burst of electricity triggers the closing response again and again. These spikes are called action potentials. And what we're seeing is the flow of charge as the membrane depolarises – the same flow of charge we see in the paramecia.

**RIGHT:** Time-lapse image of a mimosa. When touched, the leaves change from an open and rigid structure to a tightly folded appearance in a matter of seconds. If left untouched, the leaves slowly return to their original shape.

# THE UNIVERSAL NATURE OF SENSING

Few experiments in history have touched the soul of science so profoundly as the tests performed by Luigi Galvani in the latter half of the eighteenth century. Galvani was a doctor and physicist working in the academically famed town of Bologna when he stumbled across a phenomenon that would transform our understanding of life and influence the moral compass of science for the next two and a half centuries. Many different versions of Galvani's story of discovery exist, from accidental breakthrough to an inspirational leap of mind. What we do know for certain is that at some point in 1771 Galvani was in his laboratory exploring the anatomy and physiology of the frog when he noticed a remarkable effect. Whether it was Galvani or his assistant, a scalpel or a hook, at some point a statically charged piece of metal touched the exposed sciatic nerve of the dissected frog and sparked

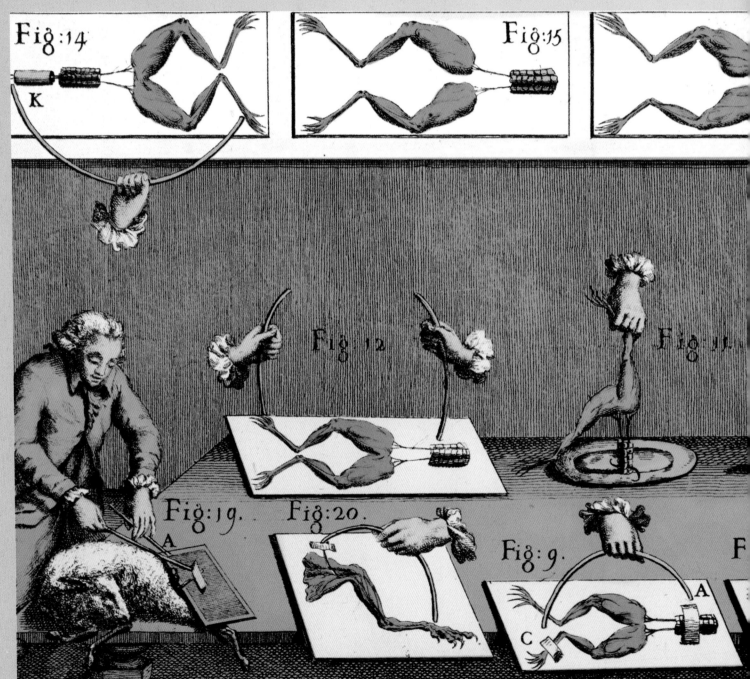

it into apparent life. It is difficult to imagine the shock that such a discovery must have had first on Galvani and then on all those who observed the phenomenon as it was repeated around the world.

At the most basic level the ability of an organism to sense its environment is based on it being able to transform an external stimulus – be it chemical, touch, sound or light – into a change in the membrane potential of a cell – an action potential. The consequences of this change depend very much on the organism in question. In a simple organism such as a paramecium the effect is very direct; changes in the concentration of ions inside the cell, regulated and shaped by the voltage and concentration-gated ion channels in the membrane, result in an immediate change in behaviour – the reversal in the direction of the beating cilia. But in other organisms

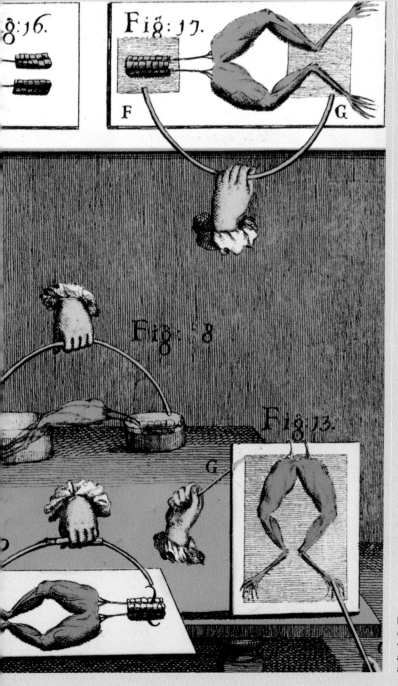

the first point of contact with the outside world can be the trigger for a whole cascade of events. In humans, our senses of touch and hearing generate action potentials directly in a similar way to the paramecium. We will explore the human ear in much more detail later in this chapter. The rods and cones in our eyes, on the other hand, do not generate action potentials directly, but ultimately stimulate them in the nerve cells that transmit the visual information to the brain. But at some stage in the transmission of signals from the senses to the central nervous system, in virtually every animal on the planet, action potentials are generated. There are always exceptions in biology, of course, and it seems that sponges are the honourable one in this case. Action potentials are also seen in plants and algae, suggesting that their use must confer advantages over other methods of signalling. The most obvious advantages are the speed of transmission and the controlled nature of the pulse; action potentials can travel along nerve cells at over 100 m/s, and the shape and intensity of the pulse itself does not change over long transmission distances. Action potentials, in other words, are a fast and reliable means of transmitting information around an organism. Their evolutionary origins probably lie in the ion channels, which are certainly very ancient. Both bacteria and algae have stretch-sensitive ion channels, and a gene has even been found in a virus that codes for a potassium channel. It is likely, then, that the basis of the senses began with the manipulation of the concentrations of ions inside and outside cells, which led to primitive sensory responses such as those seen in the paramecium. And billions of years later, this is still the way life brings the outside world in, by turning the physical world into electrical impulses. Every sensation you have, every touch, taste, sight, smell and sound, is ultimately relayed to your brain by action potentials, generated and regulated by these ancient ion channels.

Among all the diverse forms of life on Earth today, from animals, plants and protists, to bacteria and algae, the basic mechanisms used to detect the outside world are the same; they all involve ion channels in either the direct act of sensing, or the transmission of sensory data around the organism, unless you are a sponge. Yet the way in which these senses have evolved, and the stimuli they can detect, varies enormously. As we have already noted, the particular suite of senses an animal possesses is ultimately determined by its environment and lifestyle. To illustrate this, we wanted to find a large creature that experiences the world in a very different way to ourselves. We found a quite magnificent example in America's Deep South. ◉

**LEFT:** Illustration showing part of Luigi Galvani's 1771 experiment with frogs' legs, which jumped into apparent life when touched with a statically charged piece of metal.

# RIVER MONSTERS

If any one country reflects the vast range of habitats on Earth it is the United States of America. With a surface area of almost 6.5 million sq km, to travel through this country is to journey through a bewildering diversity of environments: from the tropics of Florida, to the near arctic conditions of Alaska; from the volcanic extremes of Hawaii, to the sweeping grasslands of the Midwest and the soaring mountains of Colorado. This is a land that reveals the depth of environmental diversity that Earth has to offer – countless corners providing opportunities for life to adapt and spread in ever more varied shapes and sizes. It is no wonder that the United States is one of the most ecologically diverse countries on Earth, with almost 20,000 species of plant, 90,000 species of insect, and many hundreds of different mammals, birds and reptiles all living in their own environmental niches. This diversity isn't only reflected in the size, shape and form of the animals, however; the senses of each creature are also matched to their environment, finely tuned by natural selection to allow them to survive and flourish. To appreciate how life plugs into the Universe, we have to understand not just the senses but how those senses sit within the landscape of an individual's life.

**TOP:** The giant flathead catfish is common in the Mississippi river system and can grow to an enormous size.

**ABOVE:** Don Jackson helps Brian catch a catfish with a very large net. Another, more risky, way of catching one is by the method known as 'noodling' – sticking one's hand down a catfish hole and hoping that the fish swallows it.

The landscape of America's Deep South is dominated by the great Mississippi, the largest river system in North America. The Big River rises in northern Minnesota, winding its way southwards for 4,000 km through Wisconsin, Iowa, Illinois, Missouri, Kentucky, Tennessee, Arkansas and Mississippi before emptying into the Gulf of Mexico in New Orleans. We're filming on one of its many tributaries, the Big Black River in the state of Mississippi. This is bayou country, an almost prehistoric terrain dominated by overgrown rivers and swamps. The animal life, too, has an ancient and threatening air; alligators, snakes and leeches make the river banks somewhat of a lottery, and I am delighted to remain in the safety of our boat. This is not the America of non-scalding coffee and the abjuration of personal responsibility. It is untamed, un-sanitised and exhilarating, and you can't take legal action against an alligator snapping turtle (*Macrochelys temminckii*) because it's difficult to turn the key in your SUV with a missing finger.

We are here to film one of the region's most iconic animals. The giant flathead catfish (*Pylodictis olivaris*) is a common predator in the Mississippi river system. These creatures can live for up to 20 years, and grow so large that they can approach a small adult human in size and weight. Despite their size, or perhaps because of it, the younger and more intrepid anglers in the region have developed a unique way of catching these monsters, known as 'noodling'. This involves sticking your hand into a catfish hole on the river bank and hoping that the monster decides to swallow it. Once latched on, you simply retract your arm, which is easier said than done with a hefty predator attached to your fist. Unfortunately, catfish holes are sometimes inhabited by alligators, snapping turtles or snakes, but that's extreme angling for you.

I took the decision to use nets, and enlisted the help of a quite wonderful academic from Mississippi State University, Don Jackson, who had been studying and protecting the wildlife in these river systems for longer than he cared to admit. Don laid the nets along the river the day before we arrived, and dredging them to the surface was violent and dynamic. Catfish are powerful animals, and unsurprisingly react badly to being pulled out of the water. The experience is all the more tense because it is impossible to see what is in the nets before they are hauled into the boat. The Big Black River is laden with mud and silt, turning the waters an impenetrable dirty brown. Visibility below the surface is virtually zero; you will see in the film that we tried to insert underwater cameras into the submerged nets, but, even in such a confined space, we saw virtually nothing. This is the environment in which the catfish lives and hunts, and its senses are highly specialised to allow it to catch its prey efficiently in near-zero visibility. Since its eyes are virtually useless, they are small, and its visual processing power is modest. Instead, this creature surveys its domain with a formidable array of other senses tuned by, and allowing it to dominate, its world.

Catfish get their name from the whiskery growths on their chin, which are known as barbels. These distinctive features are not for decoration – they are crucial to the animal's perception of the world. The flathead catfish has four pairs of barbels: nasal, maxillary (relating to the jawbone) and a double pair on the chin. Each of these is a multi-sensory organ that helps the fish build a detailed picture of its local physical environment. They are used to pick up the slightest vibrations in the mud of the river bed, allowing the catfish to 'hear' the tiniest movement from anything that might make a meal. They are also covered in taste buds – chemoreceptors – that tune into the chemical signals in the water. But the barbels are only the first of the many weapons in the catfish's sensory arsenal.

Sound waves travel particularly well under water and so many fish have good auditory abilities, but the catfish pushes this further than most. Along its lateral side are small pores within which sit microscopic hair cells that are extremely sensitive to low-frequency sound waves, allowing it to detect prey, lurking predators and even the occasional noodler. In common with several other species of fish, the catfish even presses its swim bladder into action as a sensory organ. The swim bladder controls the fish's buoyancy, but it is also connected to a series of small bones known as the Weberian apparatus, which further amplify vibrations in the swim bladder and pass them through to the hearing centres in the animal's head. This enables the catfish to detect higher-frequency sound waves than many other freshwater fish – a distinct advantage for a hunter.

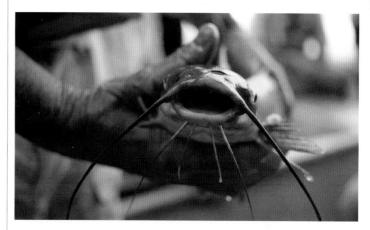

**TOP:** Catfish have four pairs of whiskers, known as barbels. These multi-sensory organs help the fish build up a 'picture' of its surroundings.

**ABOVE:** Small pores along the sides of a catfish contain microscopic hair cells that are extremely sensitive to low-frequency sound waves.

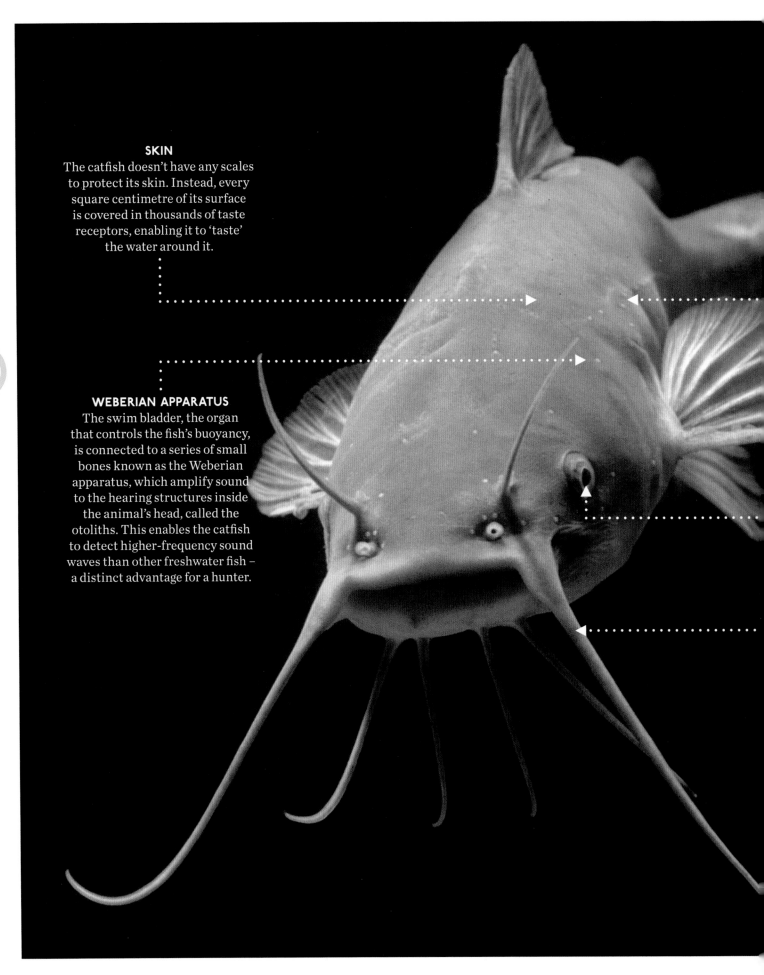

### SKIN
The catfish doesn't have any scales to protect its skin. Instead, every square centimetre of its surface is covered in thousands of taste receptors, enabling it to 'taste' the water around it.

### WEBERIAN APPARATUS
The swim bladder, the organ that controls the fish's buoyancy, is connected to a series of small bones known as the Weberian apparatus, which amplify sound to the hearing structures inside the animal's head, called the otoliths. This enables the catfish to detect higher-frequency sound waves than other freshwater fish – a distinct advantage for a hunter.

182

**HAIR CELLS**
Along the catfish's lateral side are small pores within which sit microscopic hair cells that are extremely sensitive to low-frequency sound waves, allowing it to detect potential prey and lurking predators.

**EYES**
The water is muddy, so the catfish's eyes are small and its visual processing power is modest.

**BARBELS**
Four pairs of barbels: nasal, maxillary and a double pair on its chin can detect the physical environment around them and can also pick up slight vibrations in the mud, allowing the catfish to 'hear' the tiniest movement from anything that might make a meal. They are also covered in taste buds (chemoreceptors) that tune into the chemical signals in the water.

Beyond its array of finely tuned auditory senses, the catfish possesses an ability we don't share. It can detect the electrical activity of the nervous systems and musculature of other animals. It has electro-sensing organs that are grouped together in tiny pits along its head, and, by detecting the action potentials generated within other organisms, the catfish can sense potential prey even if that prey is completely still and hidden from view.

The catfish's dominant sense, however, is taste. As Don described it to me, catfish are effectively large swimming tongues, despite the fact that they have no tongue inside their mouths. Catfish don't have scales to protect their skin; instead every available bit of their surface is covered in thousands of taste receptors. Combined with a keen olfactory system, it is able to detect thousands of different molecules in the river in concentrations of less than one part per 10 billion.

By measuring the variation in concentration of the chemicals along the length of its body, it is even able to estimate the direction of the source of the smells. In this way, the catfish builds an intricate, three-dimensional chemical picture of its world. This is something we can hardly imagine. We are dominated by our sense of sight, which allows us to build a representation of the size, position and character of the objects that surround us using light. This is because we are immersed in transparent air, and the structures we live among are illuminated by the Sun. The catfish's universe is the water and mud, populated by an array of living things leaving trails of chemicals as they move around, forming a swirling world of different tastes and concentrations, flavours and gradients. It possesses the senses it needs to navigate and hunt in this dark, dense chemical environment.

To us, the sensory suite of the catfish might seem idiosyncratic, but it illustrates an important point; no two species experience the world in the same way. Each one possesses a unique set of senses that enable it to survive and flourish in its particular environment. An animal's picture of the world is therefore both selective and subjective; it doesn't have to be complete.

One shouldn't infer from this exquisite fine-tuning that an animal's senses are in some way optimally constructed, however. The catfish is a very good example of an animal with a rather eccentric and diverse collection of sensory abilities, but starting with a blank sheet of paper, a catfish designer probably wouldn't press its swim bladder into use as a kind of internal ear. This is illustrative of a point we made earlier in this chapter. The form and function of animals can only be understood in the context of their history. Swim bladders would have come first, and since they were also good at amplifying vibrations in the water, they became coupled into the sensory system. Evolution is sometimes less of a watchmaker than a tinkering odd-job man with oily overalls and a dirty face, mixing and matching as best it can to get a job done by modifying the available parts. Nowhere is this convoluted and winding road more apparent than in the mechanics of one of our most treasured senses – the auditory system – and in the intricate but, to be frank, rather home-made apparatus involved in creating the delicate sensations of every sound we hear. ◉

## GOOD VIBRATIONS

The sensing of light by photons exciting a pigment which excites a cell potential is very biochemical. The sensing of sound depends on the sensing of a mechanical action, but has become so highly evolved that it has become so highly-tuned in some animals that they depend on hearing more than vision to 'see'.

Hearing is a mechanical sense. We humans associate hearing with the detection of sound waves in air, because we are surrounded by it. The sounds we hear are pressure waves, a travelling compression and rarefaction pattern which moves through air at 20°C at sea level at a speed of 1,225 km/h. An easy way to picture a sound wave is to look at a speaker while the music is playing. The surface of the speaker oscillates in and out, alternately increasing and decreasing the air pressure in front of it. The average young human being can detect frequencies as low as 20 Hz, which means the speaker moves in and out 20 times every second, and as high as 20 kHz, which is 20,000 oscillations per second. The sound pressure level, the difference between the pressure in the sound wave and the average air pressure in the room, determines the volume. A microphone, or an ear, works like a speaker in reverse; rather than the speaker shaking the air molecules to create a sound wave, the shaking air molecules can also cause a thin membrane to vibrate, converting the sound wave back into mechanical movement. But there is a host of engineering issues to be overcome to achieve this seemingly simple task efficiently.

**BELOW:** A fossil of a eurypterid, an extinct group of arthropods related to arachnids.

**BOTTOM:** Scorpions, such as this desert hairy scorpion (*Hadrurus arizonensis*) – here photographed under UV light – are one of evolution's great survivors.

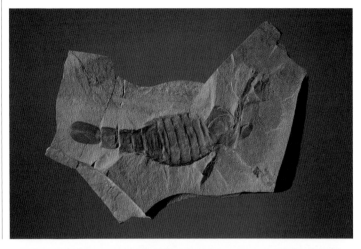

# THE POWER OF HEARING

The hearing range of animals varies from the very low (infrasound) to the very high (ultrasound) – although direct comparison of sensitivity is very difficult to measure exactly. Most hearing ranges of animals are similar, with various fine tuning due to habit and habitat. A few species – notably dolphins and bats – have developed hearing as a means of 'seeing'. This echolocation is one of the reasons that they have adapted to hear and produce sounds in very high frequencies which allows them to identify very small objects.

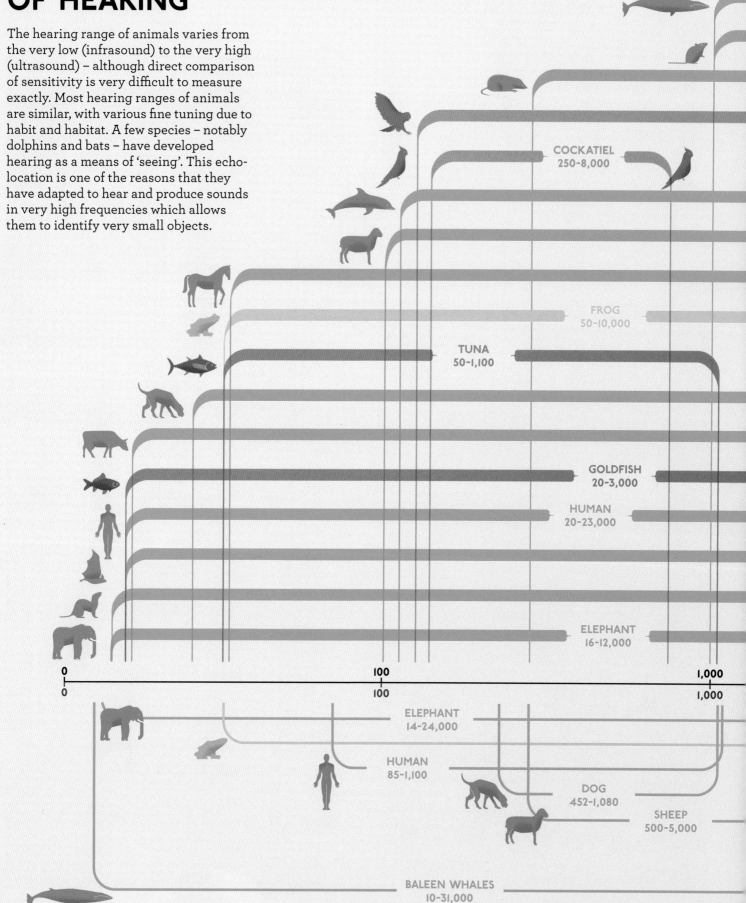

COCKATIEL
250–8,000

FROG
50–10,000

TUNA
50–1,100

GOLDFISH
20–3,000

HUMAN
20–23,000

ELEPHANT
16–12,000

0     100     1,000
0     100     1,000

ELEPHANT
14–24,000

HUMAN
85–1,100

DOG
452–1,080

SHEEP
500–5,000

BALEEN WHALES
10–31,000

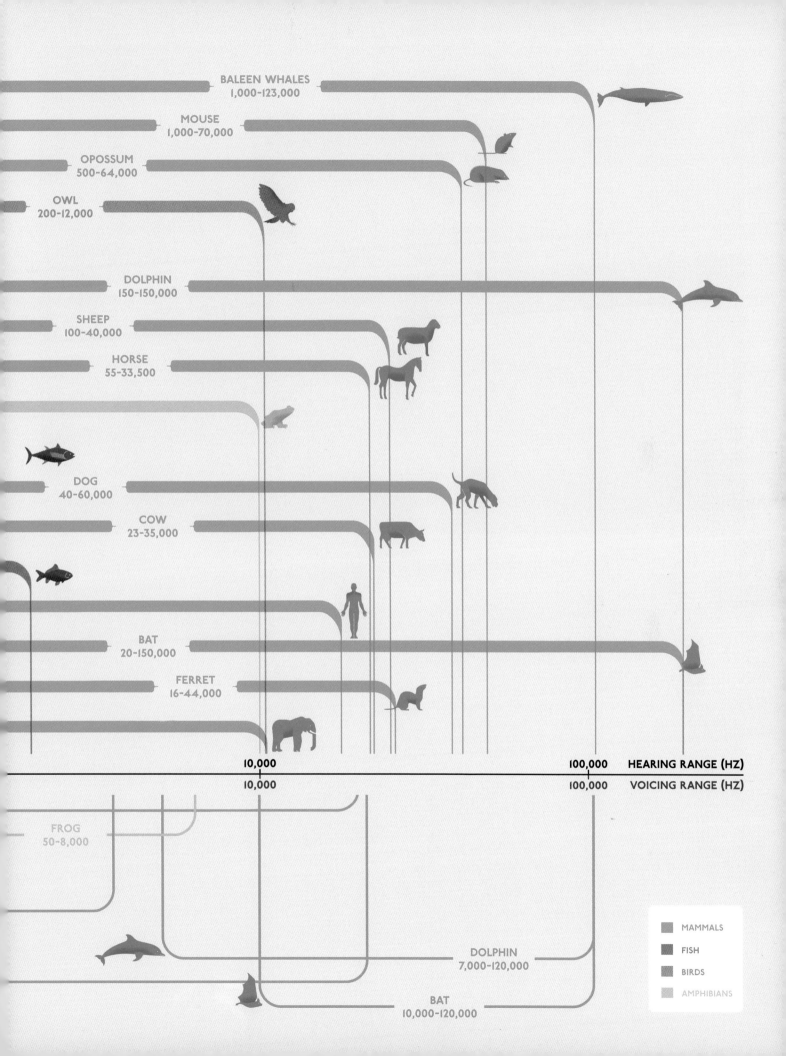

BALEEN WHALES
1,000-123,000

MOUSE
1,000-70,000

OPOSSUM
500-64,000

OWL
200-12,000

DOLPHIN
150-150,000

SHEEP
100-40,000

HORSE
55-33,500

DOG
40-60,000

COW
23-35,000

BAT
20-150,000

FERRET
16-44,000

10,000　　　　　　　　　　　　　　100,000　　HEARING RANGE (HZ)

10,000　　　　　　　　　　　　　　100,000　　VOICING RANGE (HZ)

FROG
50-8,000

DOLPHIN
7,000-120,000

BAT
10,000-120,000

MAMMALS

FISH

BIRDS

AMPHIBIANS

To search for the origins of hearing, we returned to one of my favourite filming destinations; the Mohave Desert in the Western United States. We were looking for scorpions, animals with a fearsome reputation that is largely undeserved; only around 2 per cent of species have a potentially deadly sting, and fortunately the sand scorpions of the Mohave can inflict no more damage than a bee.

For an animal renowned for living in deserts, the evolutionary history of these arid predators is surprisingly far from dry. Scorpions are one of evolution's great survivors, in the sense that they are primitive in body form and are thought to have changed little since they emerged around 450 million years ago as aquatic predators, flourishing in the shallow tropical waters of the Silurian and Devonian seas. Some of the extinct aquatic scorpions were significantly larger than their modern land-based cousins. The Pterygotus, a giant sea scorpion, could grow to over 2.1 m in length and was one of the top predators in shallow Devonian coastal waters and estuaries. But by 250 million years ago, the sea scorpions had become extinct, and it is thought that only a very small number of species, possibly only one, made the transition to the land, found a successful niche predating on other arthropods and insects and radiated into the thousand or more species we see today. Part of the key to their success is the scorpions' ability to flourish in some of the most hostile environments on Earth, something they achieved with remarkably little adaptation. The fossil record is rich with

*Tiny vibrations can be detected by the delicate hairs on the scorpion's legs and other sensory structures on its underbelly. It is a system of incredible sensitivity – so sensitive, in fact, that if a single grain of sand moves within 20 cm of the scorpion it will detect the vibration through the tips of its legs.*

**LEFT:** The desert hairy scorpion is the largest in North America, growing to 14 cm long. When it hunts, it rests its legs in a circular array around its body.

**RIGHT AND BELOW:** At the tip of each leg, the scorpion has acceleration sensors called slit sensilla, which detect low-velocity surface waves (Reynolds waves) and which are used to calculate the distance and direction of its prey.

scorpions, and reveals the detailed shifts in their anatomy necessary to make the transition. The most obvious are a thickening of the legs, the modification of their abdominal gills into book lungs, and the emergence of a preoral chamber in the mouth necessary for feeding in air. With these minor alterations, the scorpions are masters of the desert, able to survive for over a year without food or water. Living in intricately dug burrows, these nocturnal creatures emerge at night to hunt down their prey of insects, spiders and even the occasional small lizard or mouse. Like all successful predators, the scorpion needs to gather high-precision information about its surroundings to succeed in the hunt for prey, which naturally tries its best to avoid being eaten. It does this primarily by detecting sound waves in the sand.

When the scorpion is hunting, it rests its legs in a circular array around its body. At the tips of its legs, there are eight acceleration sensors called slit sensilla that allow it to detect the low-velocity surface waves (known as Rayleigh waves) created by the movement of a single sand grain at distances of up to 20 cm. By timing the arrival of these waves at each of its circularly arranged legs, the scorpion is able to calculate the distance and direction of its prey. The scorpion is, in a very effective and yet primitive sense, listening to vibrations in the sand; it can hear using its legs. Slit sensilla are unique to arachnids, where they usually serve as positional sensors, relaying information about the position of the animal's legs and changes in their orientation during

motion, via action potentials, to the animal's central nervous system. In the scorpion, some appear to have been co-opted into a basic but effective auditory system, adapted to detect slow-moving Rayleigh waves travelling on the surface of the sand, and thereby allowing this unique animal to become a deadly predator in the harsh desert environments in which it has found a unique niche.

This ability to map sound waves onto position is not unique to scorpions. Although the details are different, a similar principle is used inside our ears. Sound waves are translated into mechanical vibrations, in our case by a specialised structure in our ears called the basilar membrane. The vibrating membrane disturbs hair cells, and this opens ion channels, leading to membrane depolarisation and ultimately action potentials travelling down the spiral ganglion neurons to the brain. The language we used in the above paragraph is admittedly quite technical, but hopefully you recognise the terms and concepts because we've met them all before in this chapter. Again and again, from the paramecium to the scorpion to the human being, the basic underlying mechanisms of the senses are all very similar. So let us dig a little deeper into the mechanics of the human ear, because it provides a quite spectacular insight into the mechanisms behind, and the evolutionary origins of, hearing. It also serves to illuminate our own journey, mirroring that of the scorpion, from the Devonian oceans to the modern-day land. ◉

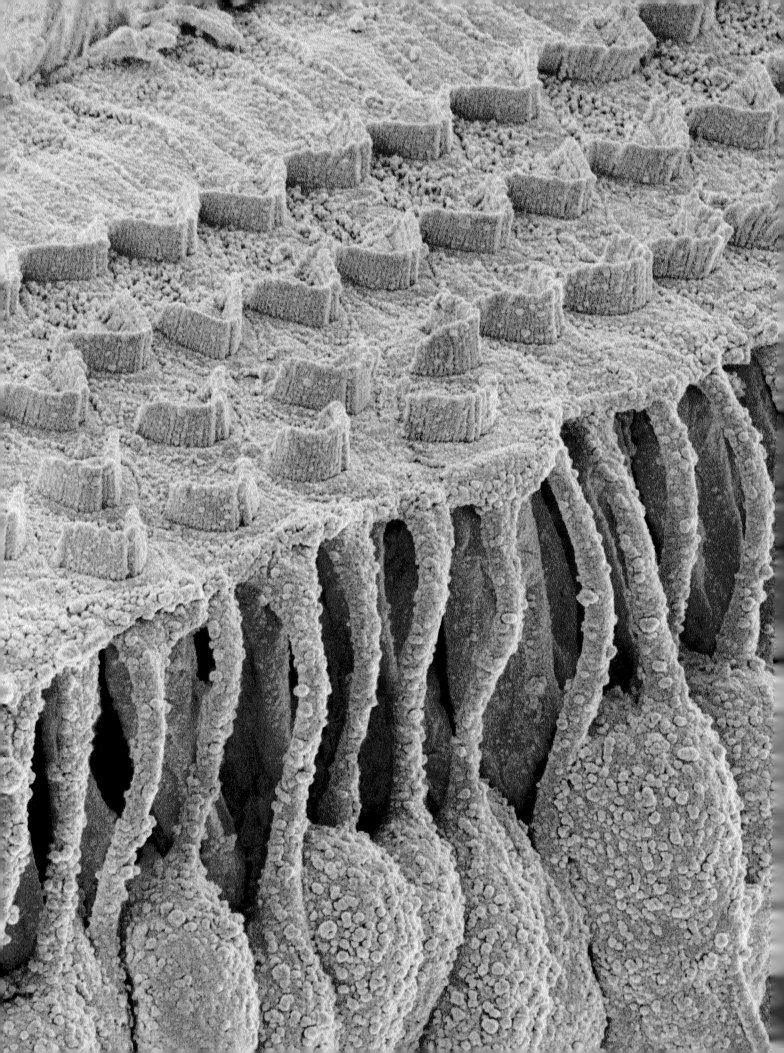

# THE HUMAN EAR: A WONDER OF ACOUSTIC ENGINEERING

**LEFT:** The sensory hair cells in the cochlea, the inner ear's auditory sense organ. The crescent-shaped hairs at the top are the sterocilia, which bend when sound waves enter the inner ear, triggering the release of neurotransmitter chemicals that generate nerve impulses.

Our ears are complex pieces of machinery because they have to perform a complex task. They have to turn sound waves into electrical signals – action potentials – and deliver them to the brain for processing. And they have to do this over a wide range of frequencies and amplitudes. We can hear the quiet buzz of a mosquito's wings, and yet sit within a few metres of a roaring riverboat engine without incurring hearing damage, even though the power delivered into our ears may be a hundred million times greater. In all land vertebrates, the conversion of sound waves into electrical signals is carried out by a vibrating membrane called the basilar membrane. This membrane sits deep inside our inner ear in a spiral structure called the cochlea (see diagram, p 192), and varies in thickness, width and mass along its length. At the entrance to the cochlea, it is taught and thin, while at the far end, it is 'floppier', but

*At the entrance, or base, of the cochlea, the basilar membrane vibrates in response to high frequencies, while at the far end, or apex, the basilar membrane responds to low frequencies.*

wider and of a higher mass. This arrangement means that different frequencies of sound cause the basilar membrane to vibrate with maximum amplitude at different positions along its length. At the entrance, or base, of the cochlea, the basilar membrane vibrates in response to high frequencies, while at the far end, or apex, the basilar membrane responds to low frequencies. Hairs attached to the basilar membrane, called stereocilia, pick up these vibrations, and through the now-familiar process of the opening of ion channels and the depolarisation of cell membranes, action potentials travel along the spiral ganglion nerves to the brain. This is an extremely clever piece of engineering, translating different frequencies of sound into a position measurement, which is then interpreted by the brain.

There is a fundamental problem to be overcome, however, before this elegant piece of machinery can do its job. The solution is fascinating, because it is one of the most spectacular examples of our evolutionary past being written into the very anatomical structure of our senses. The problem is this; although we live in air, our bodies are filled with water. In particular, the cochlea is filled with fluid. Sound does not travel at all well from air into water, as you will know if you are a swimmer. If you dive down to the bottom of a swimming pool, you can hear virtually nothing happening above the surface. This is because the water surface is a near-perfect reflector of sound waves; over 99.9 per cent of the power bounces back and virtually none is transmitted into the water. An engineer would call this an impedance matching problem, and it requires a mechanical solution. Our ocean-dwelling ancestors didn't face this problem of course; as we saw in the catfish, sound waves travelling in the water will happily pass into the watery body of the fish and vibrate its swim bladder without any need for complex mechanics. But as soon as the first animals emerged from water into air, hearing became a challenge, and in odd-job tinkering mode, evolution came up with an ingenious solution. ◉

# THE OSSICLES: ONE OF NATURE'S GREAT EVOLUTIONARY BODGES

The tricky job of getting sound waves into the fluid-filled cochlea is carried out by the three smallest bones in the human body; the malleus, the incus and the stapes. Known collectively as the ossicles, these bones sit in the middle ear, connecting the ear drum to the oval window at the entrance to the cochlea. Their Latin names describe their shape; the malleus resembles a mallet, the incus looks like an anvil, and the stapes is stirrup-shaped. The malleus is the largest, and attaches to the eardrum. It is connected via the incus to the stapes, which in turn is attached to the oval window. There are two features of this arrangement which allow for the efficient transmission of sound from the eardrum into the inner ear. Firstly, they act like a level system, magnifying the movement of the eardrum. Secondly, the footprint of the stapes is 17 times smaller than the area of the eardrum, which means that the vibrations of the eardrum are transmitted into the fluid of the inner ear with much greater force. The result is that, rather than 99.9 per cent of the sound energy being reflected at the interface between air and fluid, around 60 per cent is transmitted into the inner ear. These three little bones do an excellent job! But how could this wonderful piece of acoustic engineering have evolved? The story begins in the water, and isn't entirely historic, because there are still animals alive today that use different forms of these same three bones for very different purposes. ◉

## ANATOMY OF THE HUMAN EAR

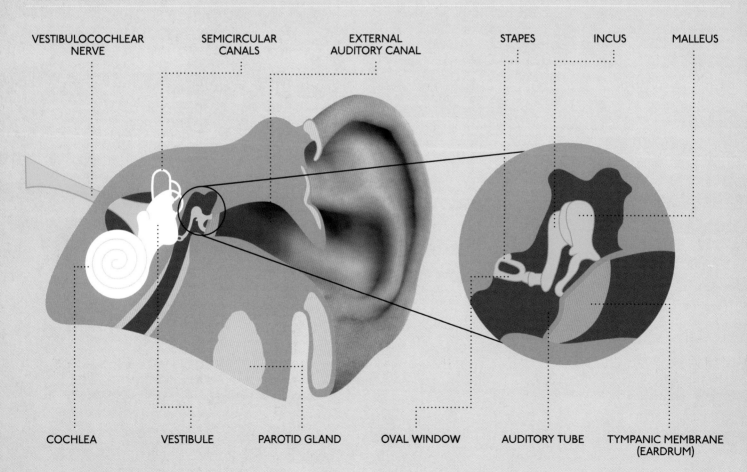

VESTIBULOCOCHLEAR NERVE

SEMICIRCULAR CANALS

EXTERNAL AUDITORY CANAL

STAPES

INCUS

MALLEUS

COCHLEA

VESTIBULE

PAROTID GLAND

OVAL WINDOW

AUDITORY TUBE

TYMPANIC MEMBRANE (EARDRUM)

BELOW: The eardrum, or tympanic membrane, is connected to the inner ear via the three smallest bones in the human body: the malleus, the incus (both shown here) and the stapes (see opposite).

INCUS

MALLEUS

STAPES

EARDRUM

194

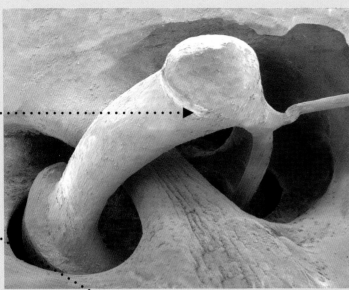

**LEFT:** The stapes (shaped like a tiny stirrup: 'stapes' means 'stirrup' in Latin) transmits vibrations to the fluid-filled cochlea of the inner ear, where they are converted to nerve impulses.

**BELOW:** The incus (top left) transmits vibrations to the stapes (pink).

# EVOLVING EARS AND EYES

The mammalian ear did not proceed in a linear progression. The middle ear bones evolved from the jawbones of our reptilian forebears, with the bones gradually being repurposed for the sake of improved hearing, which in turn occurred parallel to the refinement of mammalian teeth.

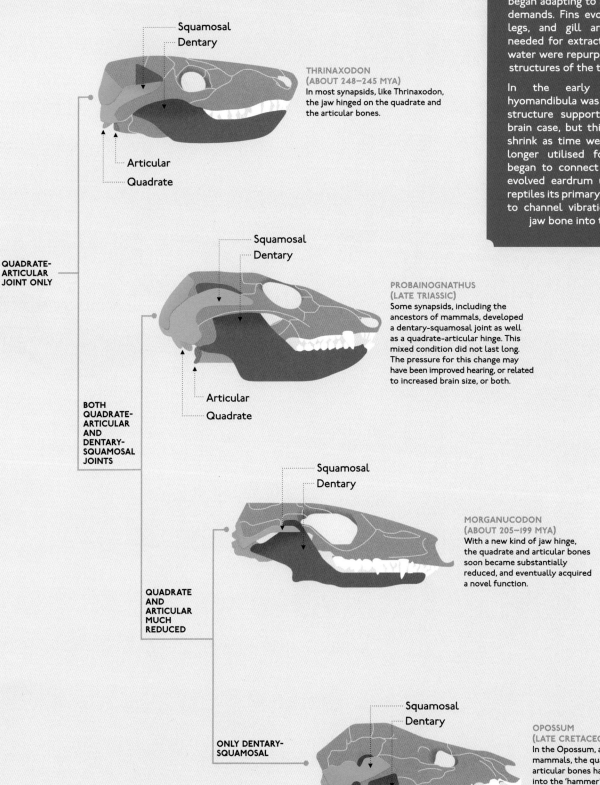

**Squamosal**

**Dentary**

**THRINAXODON**
**(ABOUT 248–245 MYA)**
In most synapsids, like Thrinaxodon, the jaw hinged on the quadrate and the articular bones.

**Articular**

**Quadrate**

**QUADRATE-ARTICULAR JOINT ONLY**

**Squamosal**

**Dentary**

**PROBAINOGNATHUS**
**(LATE TRIASSIC)**
Some synapsids, including the ancestors of mammals, developed a dentary-squamosal joint as well as a quadrate-articular hinge. This mixed condition did not last long. The pressure for this change may have been improved hearing, or related to increased brain size, or both.

**Articular**

**Quadrate**

**BOTH QUADRATE-ARTICULAR AND DENTARY-SQUAMOSAL JOINTS**

**Squamosal**

**Dentary**

**MORGANUCODON**
**(ABOUT 205–199 MYA)**
With a new kind of jaw hinge, the quadrate and articular bones soon became substantially reduced, and eventually acquired a novel function.

**QUADRATE AND ARTICULAR MUCH REDUCED**

**Squamosal**

**Dentary**

**OPOSSUM**
**(LATE CRETACEOUS–RECENT)**
In the Opossum, and all living mammals, the quadrate and the articular bones have been transformed into the 'hammer' and the 'anvil' of the inner ear. This transformation occurred separately in the two lineages of mammals – the monotremes (platypus and echidnas) and the eutherians (placental mammals).

**ONLY DENTARY-SQUAMOSAL**

# EYE EVOLUTION IN LIVING MOLLUSCS

Certain components of the eye appear to have a common ancestry. However, complex, image-forming eyes evolved dozens of times. Multiple eye types and subtypes developed simultaneously in various animals, showing a wide range of adaptations. The different forms of eye in molluscs are often cited as examples of parallel evolution to the development of the human eye.

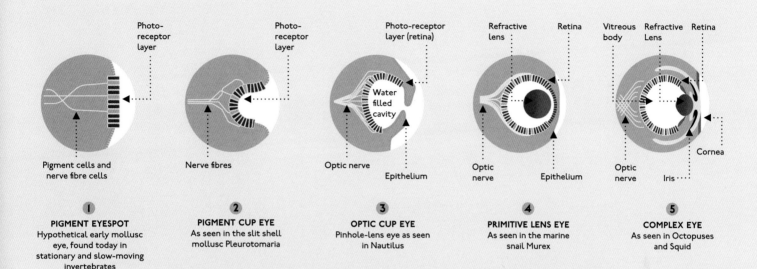

Photo-receptor layer

Pigment cells and nerve fibre cells

**① PIGMENT EYESPOT**
Hypothetical early mollusc eye, found today in stationary and slow-moving invertebrates

Photo-receptor layer

Nerve fibres

**② PIGMENT CUP EYE**
As seen in the slit shell mollusc Pleurotomaria

Photo-receptor layer (retina)

Water filled cavity

Optic nerve

Epithelium

**③ OPTIC CUP EYE**
Pinhole-lens eye as seen in Nautilus

Refractive lens

Retina

Optic nerve

Epithelium

**④ PRIMITIVE LENS EYE**
As seen in the marine snail Murex

Vitreous body

Refractive Lens

Retina

Cornea

Optic nerve

Iris

**⑤ COMPLEX EYE**
As seen in Octopuses and Squid

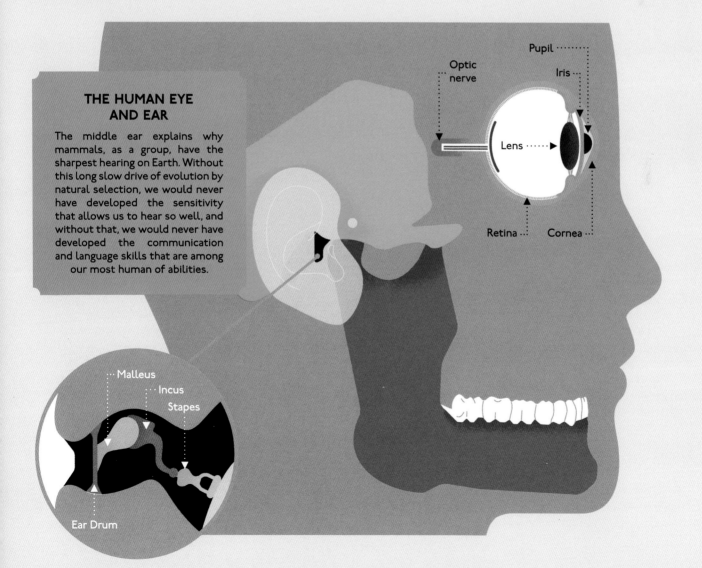

## THE HUMAN EYE AND EAR

The middle ear explains why mammals, as a group, have the sharpest hearing on Earth. Without this long slow drive of evolution by natural selection, we would never have developed the sensitivity that allows us to hear so well, and without that, we would never have developed the communication and language skills that are among our most human of abilities.

Optic nerve

Pupil

Iris

Lens

Retina

Cornea

Malleus

Incus

Stapes

Ear Drum

# THE JAWLESS
# LAMPREY

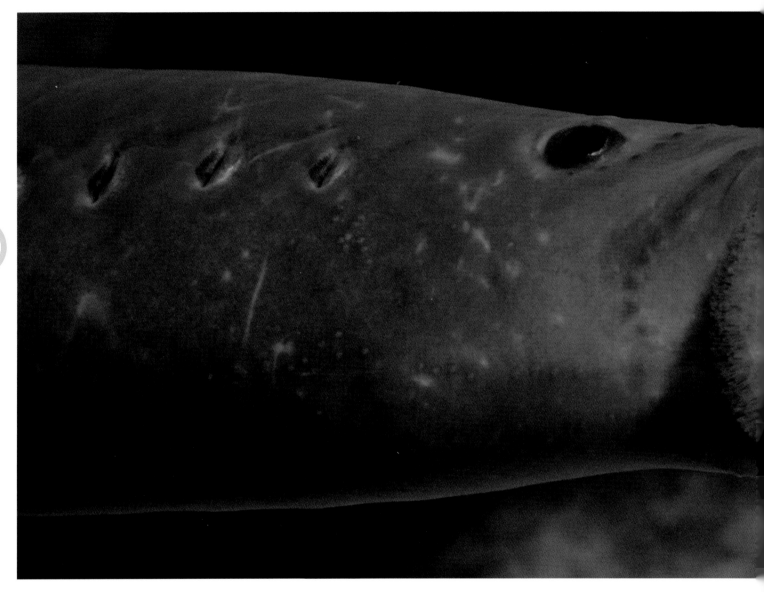

If you are looking for 'Alien' creatures on our planet, very few would fit the bill as well as the group of jawless fish known as the Agnathans, which include the modern-day lamprey. These peculiar-looking creatures are often mistaken for eels, but they are fish with big eyes, a single nostril and strange, prehistoric-looking jawless mouthparts. Continuing the alien theme, there are species of lamprey that feed by attaching their mouth to their prey and grating their way through the flesh using their multitude of teeth until they reach the blood and bodily fluids within. They have even been known, albeit rarely, to attack humans if they are very hungry. As well as being unarguably odd-looking creatures, the jawless lamprey also provide a window back in time, to a world where evolution had yet to drive the structure of the familiar-shaped jaw that we see across the animal kingdom today.

Modern Agnathans are similar to some of the earliest vertebrates that lived on Earth, a glimpse of life 500 million years ago in the Paleozoic seas, and the earliest Agnathan fossils date back to the Cambrian. To understand how a jawless fish is related to the intricate mechanics of our hearing, we need to look closely at the structure of the lamprey's head. On each side there are seven or more gill pores, whose function – as in all fish – is to extract oxygen from the water. The gills are made up of thin filaments of tissue that create a large surface area to absorb dissolved oxygen. To increase the efficiency of this process, the gills also contain a set of bony structures known as gill arches. These support the gills by separating their surfaces so that they can extract the dissolved oxygen as effectively as possible. These seemingly innocuous bones have had a huge effect on the body plan of every vertebrate on Earth.

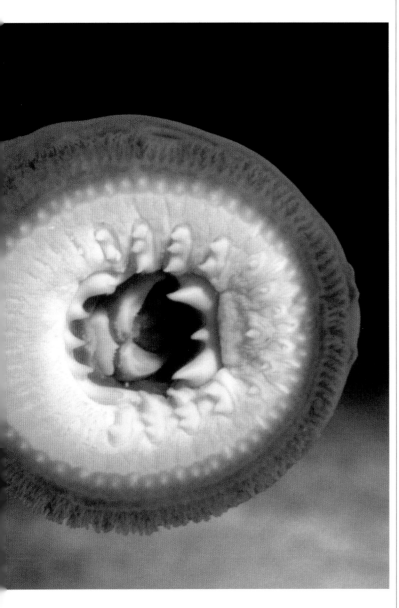

*The gills are made up of thin filaments of tissue that create a large surface area to absorb dissolved oxygen.*

Around 530 million years ago, following the Cambrian explosion, the oceans were populated by jawless fish which were also some of the first vertebrates. Over a period of around 50 million years, the front gill arches migrated forward in the head to form jaws. The fossil record contains many examples of jawed fish, the earliest being the Acanthodians, found in the late Silurian period beginning around 450 million years ago and now extinct. In modern jawed fish you can still see this mixed configuration, with the forward upper and lower jawbones, rear gill arches, and an interesting third bone, modified from the second set of gill arches, known as the hyomandibular, which supports the rear of the jaw and plays a role in helping guide water over the gills.

Around 400 million years ago, the first vertebrates made the journey from sea to land, and the now-redundant gill arches faded away and were subsumed into other structures in the head and throat. The hyomandibular shrank, and was co-opted to perform a different function. It picked up vibrations in the jaw, and channelled them into the inner ear of the newly emerged reptiles. Modern-day crocodiles and alligators, whose ancestors emerged around 320 million years ago, still use the hyomandibular in precisely this way.

Around 210 million years ago, the first mammals appeared. While the reptilian jaw is composed of several bones fused together, mammals have only single upper and lower jawbones. In mammals, two of the 'unused' bones, the articular bone from the lower jaw and the quadrate bone, which forms part of the linkage between the skull and the upper jaw in most modern reptiles, birds and amphibians, shrank away and joined the hyomandibular in the ear. The articular became the malleus, the quadrate is the incus and the hyomandibular is the stapes.

In 2007, a 125-million-year-old fossil of a small mammal known as a yanoconodon was discovered in China. The tiny animal, only around 13 cm long, had three inner ear bones similar in size and shape to those found in mammals today, but yet still connected to the jaw. The yanoconodon is an excellent example of an intermediate fossil – a snapshot of the evolutionary journey of the ossicles from gill arches to jawbones to independent, high-precision impedance matching devices in the ears of modern mammals.

The evolution of mammalian hearing is, for me, one of the most wonderful stories we filmed in *Wonders of Life*. It is a grand synthesis of ideas, bringing together the vibration-sensing capabilities of the scorpion, the physics of ion-channels and action potentials as seen in one of the most primitive of all organisms, and the ability of evolution by natural selection to shift the function of bones, over hundreds of millions of years, to solve a basic problem in acoustics. The result is the spectacular specialisation of the human ear, a device the origins of which would be absolutely baffling without a detailed understanding of its evolutionary history. Our ears are the very epitome of the idea that biological systems are inherently four-dimensional, in the sense that their deep history is literally built into their form and function. We evolved from ancient jawless fish, whose gill arches we press into service every time we engage in conversation or are distracted by a cherished song. ◉

199

# LET THERE BE LIGHT...

'...if numerous gradations from a perfect
and complex eye to one very imperfect
and simple, each grade being useful
to its possessor, can be shown to exist;
if further, the eye does vary ever so
slightly, and the variations be inherited,
which is certainly the case; and if any
variation or modification in the organ be
ever useful to an animal under changing
conditions of life, then the difficulty
of believing that a perfect and complex
eye could be formed by natural selection,
though insuperable by our imagination,
can hardly be considered real.'
*Charles Darwin*

**RIGHT:** The view from Nicéphore
Niépce's window, photographed in
1826. The pinhole-camera mechanism
used by Niépce is not too dissimilar
to our own visual system.

It may be black and white and rather hazy, but the image below is one of the most iconic in the history of photography. Created in 1826 by Nicéphore Niépce, it is the oldest surviving permanent photograph. Niépce took the image from a window in his home town of Saint-Loup-de-Varennes, over a period of eight hours, transforming an ordinary scene into a piece of history. The picture was captured using a camera obscura to focus the reflected light from the rooftops onto a light-sensitive plate. The camera obscura itself was a very ancient invention – Aristotle had used one to view a partial solar eclipse over two thousand years earlier. But what set this particular example apart was the pewter plate set in place of the viewing screen. The plate was covered with a tar-like mixture known as bitumen of Judea, which hardened when exposed to light. After his 8-hour exposure was completed, Niépce washed away the unexposed, unhardened material with a solvent, leaving an image of his window view; an unremarkable moment frozen forever.

At its most basic level, our visual system is not too dissimilar to Niépce's camera, because the behaviour of visible light determines to a large extent the form of any device designed to detect it. Our eye has a pinhole-size pupil that allows light into a darkened chamber. Where Niépce placed his pewter plate, we have a retina – a collection of light-sensitive cells that convert the image into electrical signals that are carried to the brain via action potentials in the optic nerve. With its complex lens, high-resolution imaging capabilities and superb low-light sensitivity, the human eye is of course significantly more capable than Niépce's primitive camera obscura. In fact, it is significantly more capable in many ways than the finest cameras of today. It is a masterpiece of engineering, and as such it has entered popular culture in some quarters as a supposed challenge to science; how could something so intricate have evolved? Without further ado, let us describe how the eye works, and how it evolved. ◉

# SEEING THE LIGHT

**LEFT:** False-colour scanning electron micrograph of the human retina, showing the central fovea, the crater-like depression in the photosensitive layer of the eye.

**BELOW:** False-colour scanning electron micrograph of a section through the human retina, the light-sensitive tissue that lines the inside of the eye, showing receptor cells (red), rods (white) and cones (yellow).

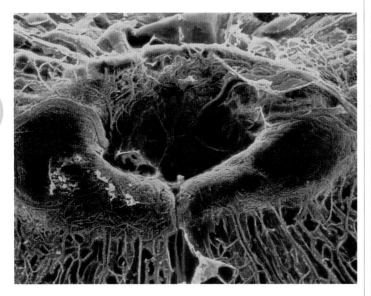

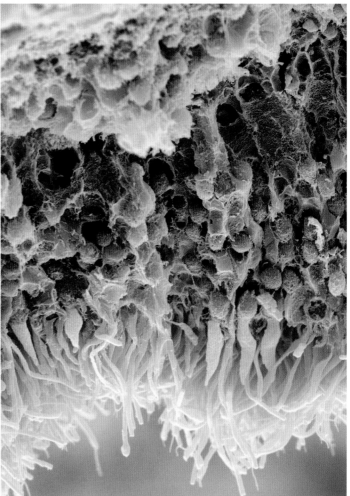

Throughout this book, we have seen that a powerful route to discovering how organisms evolved certain characteristics is to look for commonalities across radically different forms of life. We noted, for example, that all green plants and algae carry out oxygenic photosynthesis in exactly the same way, and this supported the conclusion that oxygenic photosynthesis evolved once in the cyanobacteria. Likewise, we observed that all living things on the planet use proton gradients, and this may point to a common origin around alkaline vents in the acidic oceans of the primordial Earth. In the case of vision, it would seem at first glance that there is little commonality, because there is a truly vast range of eyes in the animal world that use radically different designs to achieve the same goal: the compound eyes of a mantis shrimp are quite unlike our own. But if we consider the most basic function of any eye, which is to detect light, then we find a different story, because the underlying biochemistry of light detection turns out to be near-universal, and this is strongly suggestive of a common evolutionary origin.

The human visual system has two types of light-sensitive cells: rods and cones. The rods are more sensitive, but only allow us to see in black and white. We possess three types of cones, each of which has its peak sensitivity tuned to different wavelengths of light, giving us colour vision. All of these cells share the same underlying biochemistry. They use a pigment called retinal, a form of vitamin A, bound to proteins called opsins. Each of the opsins in these cells is slightly different, and they tune the response of the cells to particular colours. There is also a third type of photosensitive cell in our eyes, associated with our body clock. These are called photosensitive retinal ganglion cells, and, while they are different in structure to the rods and cones – a fascinating difference that we'll return to later – they too are constructed from retinal, bound to an opsin. The whole family of similar molecules is often referred to generically,

*It can take up to 20 minutes for enough rhodopsin to be re-formed to give us our full night vision, even with our mitochondria churning away at full speed.*

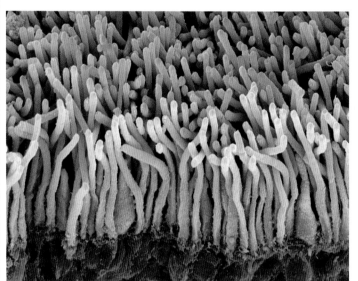

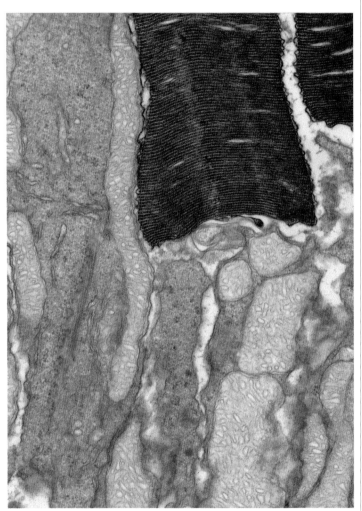

and it has to be admitted, slightly sloppily, as rhodopsin. We'll do that from now on, although it is worth bearing in mind that, strictly speaking, rhodopsin refers to the particular molecule found in rods. In the cones, the molecules are known as cone opsins, and the photosensitive retinal ganglion cells contain melanopsin.

In our eyes, therefore, everything starts with rhodopsin. Photons enter the eye through the lens and are absorbed by rhodopsin molecules. This causes a structural change in the molecules themselves, which ultimately results in an action potential being generated, which is whisked off down the

optic nerve and into the visual cortex of the brain. Then, in a relatively slow process, our old friends the mitochondria step in, releasing the energy necessary to re-form the rhodopsin molecules ready for the detection of more photons. This is why it takes a while to recover from being 'blinded' by a bright light at night. It can take up to 20 minutes for enough rhodopsin to be re-formed to give us our full night vision, even with our mitochondria churning away at full speed.

This is quite an involved mechanism, but here is the remarkable thing – it is the smoking gun that allows us to begin to unravel the story of the evolution of vision. Rhodopsin is universal. Every eye in every animal on the planet uses rhodopsin, or closely related molecules. This suggests a very ancient origin indeed, because there are eyes on our planet that have arisen quite separately from each other and are separated by great swathes of evolutionary time. The mantis shrimp is an excellent example. To find the common ancestor of mantis shrimps and humans, we have to go back at least 540 million years to the early Cambrian, and yet we share the same basic visual machinery, based on rhodopsin. But what is that common ancestor? Let's follow the timeline back, in a fashion that rather resembles a detective story. We'll be led, intriguingly, to the conclusion that rhodopsin may have been an evolutionary invention that predates all animals by a long, long time.

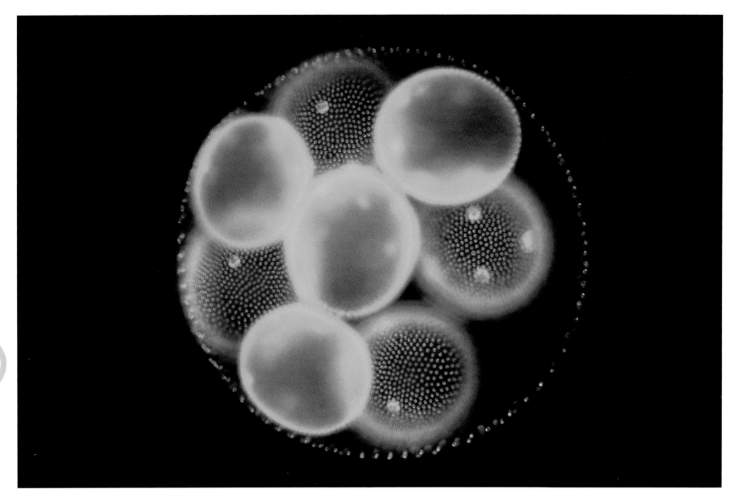

## THE ANCIENT ORIGINS OF VISION

The bright green slime called volvox is a common type of single-celled green photosynthetic algae that can be found in freshwater ponds and puddles all over the world. Common it may be, but its seemingly featureless ubiquity obscures a quite unexpected level of complexity. Volvox live as large spherical colonies. The sphere you can see in the picture above is made up of around 50,000 individual volvox cells connected together by thin strands of cytoplasm. Each individual alga has two tail-like hairs, called flagella. By synchronising their beating flagella, the whole colony can move around in a coordinated fashion. This is a beautiful example of the fine line between single-cellular and multi-cellular life. Here is a collection of individual cells working together for the good of a single multi-cellular entity – just like us.

As photosynthetic organisms, volvox use their flagella to direct the colony towards the strongest sources of light. This, of course, requires them to know where the light is. If you look closely at the image of a single volvox cell, you can see a tiny red spot. This pigmented area is an 'eye spot' – a photosensitive region that controls the beating of the flagella and allows the algal cell to swim towards the light. When the eye-spots are stimulated by bright light, they command the flagella to stop; when the light levels dim, the stop sign is lowered and the search for light begins again. So integrated are these eye spots into the behaviour of the algal colony that more of them are found in the cells on one side of the colony than the other – in effect, the sphere has a front and a back,

with the photosensitive system leading the way forward. It is an amazing level of coordination for a single-celled organism. But the clue we need in order to follow our detective story is that the microscopic 'visual' system of the volvox is based on a form of rhodopsin known as channelrhodopsin.

This is a fascinating observation whose significance is currently a matter of scientific debate. The suggestion is that the close similarities between the rhodopsins in volvox and other algae and the rhodopsins in the visual systems of all animals imply a common origin. Intriguingly, and adding to our pile of evidence, there are also strong similarities between the genes that control the emergence of the algal eye-spots and the genes that control the development of our own eyes. But vision cannot have emerged in the algae and ended up in animals, because algae are not a part of our branch of the tree of life; we are not directly related to them. So, our detective must do more work. If we are to assert a common origin of the visual sense in volvox and animals, we have to search further backwards down the tree of life.

Some scientists have suggested that, as with oxygenic photosynthesis, we might turn our attention to that ancient and successful group of organisms, the cyanobacteria. As we saw in Chapter 1, it is now widely accepted that the chloroplasts, the seat of photosynthesis in all green plants and algae, are the still-productive remains of an ancestral cyanobacterium that was engulfed by another cell. One theory for the origin of vision is that a similar thing happened to one of our very early ancestors, the protozoa. It is certainly

**LEFT:** *Volvox globator*, a type of single-celled green photosynthetic algae, live as large spherical colonies and can move around in a coordinated fashion, using the tail-like hairs, called flagella, on each individual alga.

**BOTTOM:** Volvox have an 'eye spot', shown as a tiny red dot on this image, which controls the flagella (also just visible here), allowing the algal cell to swim towards the light.

**LEFT:** *Volvox globator*, a type of single-celled green photosynthetic algae, live as large spherical colonies and can move around in a coordinated fashion, using the tail-like hairs, called flagella, on each individual alga.

**BOTTOM:** Volvox have an 'eye spot', shown as a tiny red dot on this image, which controls the flagella (also just visible here), allowing the algal cell to swim towards the light.

**RIGHT:** Computer-generated model of the internal workings of a human eye, showing the rhodopsin (blue) connected to a molecule of the light-sensitive compound retinal (yellow).

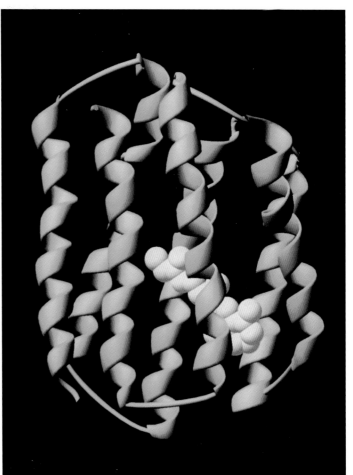

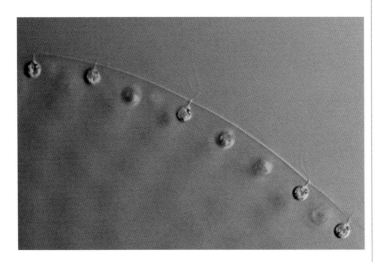

true that there are protozoa today with surprisingly complex eyes, using rhodopsin, which appear to be associated with the remains of chloroplasts. Could it be that the origin of vision was a chance event – the formation of a chimera – in a process similar to that which led to the development of oxygenic photosynthesis? The answer is that nobody is sure yet. But there is overwhelming evidence, as we've seen throughout this book, that endosymbiosis – a merger between organisms – was responsible for many of the great evolutionary leaps, including the emergence of the eukaryotic cell, oxygenic photosynthesis and, maybe, vision. The probability of these mergers is unknown; they may be virtually inevitable, given enough time and the vast numbers and fast reproduction rates of single-celled organisms, or they may be extremely rare.

This is the end of our little detective story. It is certainly true that all animals use very similar forms of rhodopsin at the heart of their visual systems. It is also true that some algae also use a different, but not too different, form of rhodopsin as very primitive light sensors. The suggestion is that this implies a common ancestor, in which rhodopsin was first co-opted into the first visual sense, and that common ancestor maybe a cyanobacterium. This cyanobacterium delivered the necessary technology to both algae and protozoa via endosymbiosis, and that is why we share rhodopsins with volvox.

For me, there is something strange and wonderful, yet slightly disconcerting, about the possibility that my sight may have its origins in a cyanobacterium. Disconcerting

might be the wrong word – I can't quite put my finger on it. Dizzying might be better. Sight seems to me to be the most direct bridge between the internal and external world. It is almost as if the external world would be an irrelevance if it went unseen. And, somewhat irrationally, I find that it is one thing to suggest that the origin of the chloroplasts in green plants and algae is the result of the merger between whole, functioning organisms, but quite another to speculate that I see the world as a result of the action of ancient bacterial genes, passed down faithfully over billions of years as a result of the chance merger between two ancient cells. If correct, this is a most visceral demonstration of the deep interconnectedness of life on Earth.

To summarise, all animals share the same basic biochemistry of vision, based on rhodopsin, and this suggests a common origin which may be extremely ancient, predating the split between animals and algae. Studies of the genes controlling the development of the visual sense givesadded weight to this theory. But rhodopsin alone does not deliver vision at the level of sophistication present in humans and higher animals. Our retina, complex as it is, requires a host of other machinery to enable it to deliver the information necessary for our brains to construct a colourful picture of the world. The emergence of this machinery is much more recent in the history of life, because while light-sensitive pigments are undoubtedly very ancient, the first evidence for complex eyes is to be found in the fossils of the Cambrian explosion, 'only' 540 million years ago. ◉

**BELOW:** The eye of a southern ground hornbill ('Bucorvus leadbeateri'), a species native to Africa.

**MIDDLE:** The light-sensitive cells of a giant clam ('Tridacna gigas') can distinguish dark and light as well as shadows.

**BOTTOM:** The compound eyes of a dragonfly contain thousands of tiny hexagonal eyes.

**BELOW:** Tarantulas have eight eyes: two wide ones at the front, four smaller ones underneath, and two more small ones on the side of the upper head.

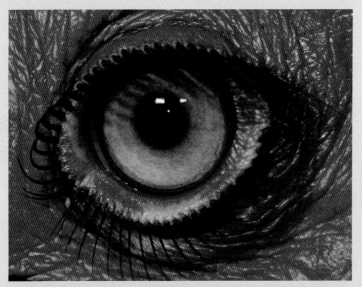

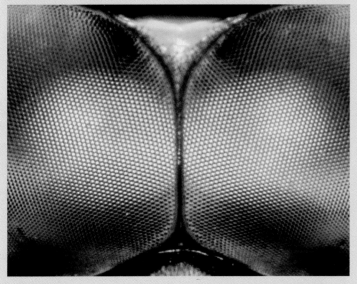

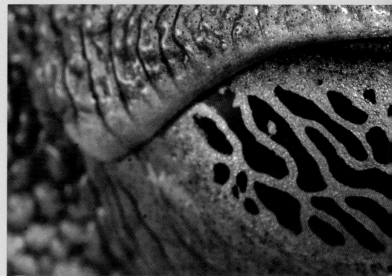

**BOTTOM:** The closed eye of a red-eyed tree frog ('Agalychnis callidryas'), showing the nictating membrane, which protects the eye.

**RIGHT:** The bulging eyes of the veiled chameleon ('Chamaeleo calyptratus') rotate independently.

**MIDDLE:** Long eyelashes curve downwards over the eye of this Asian elephant ('Elephas maximus').

**BOTTOM:** A circle of bright yellow skin surrounds the eye of this hyacinth macaw ('Anodorhynchus hyacinthinus').

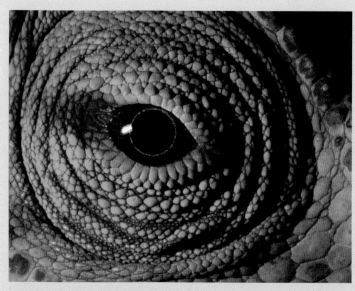

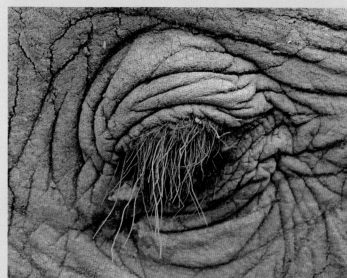

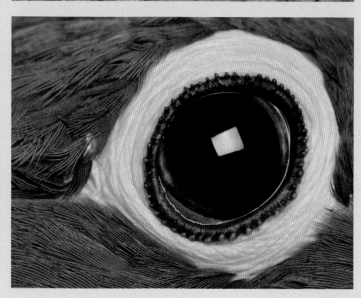

# EYE WIDE OPEN

These pictures are of a trilobite, an extinct marine arthropod and ferocious armoured predator that dominated the oceans for well over 300 million years. Trilobites appeared early in the Cambrian explosion 540 million years ago, and a tremendous number of fossils of these iconic creatures have been discovered at sites such as the Burgess Shale in Canada.

Trilobites had big eyes that dominated the head of the animal. They are compound eyes, not dissimilar to those of the mantis shrimp in basic structure. The trilobite's eyes have a surprising feature, however, shared by only one other animal. Each of the hundreds of individual lenses that make up its compound eyes is made of calcite – transparent crystals of calcium carbonate. The only living animal known to share this lens structure is a species of brittle star known as *Ophiocoma wendtii*. These starfish-like creatures are covered in tiny, high-precision calcite lenses that focus light onto their many photoreceptors. The animal, it has been remarked, is like a single compound eye. The interesting thing about the opportunistic eyes of the brittle star and the trilobite is that they have taken something 'lying around' – in this case the material of their skeletons – and co-opted it to act as a lens. This appears to be the evolutionary origin of lenses; anything lying around will do. This idea is supported by the vast array of different lenses in use today across the natural world. The proteins used in all vertebrate eye lenses, including humans, are called crystallins. Some of these proteins are active enzymes, also used around the body for other purposes. All are closely related in structure, all are found outside of the eye lens as well as inside, and only a few are common to all vertebrates.

This suggests that, just as in the trilobites and brittle stars, vertebrates used whatever they had available to form a lens. Crystallins, in other words, predate the evolution of the eye and, over time, were re-purposed and pressed into service in the visual system.

The question is, over how much time, because this seems like quite a tricky thing to accomplish. In 1994, Dan Nilsson and Susanne Pelger published an influential article named 'A pessimistic estimate of the time required for an eye to evolve', in which they calculated the number of steps necessary to go from a simple eye spot – a cluster of rhodopsin molecules formed into a flat, naked retina – to a complex camera eye. With a simple model of natural selection, and assuming one generation per year, they showed that a camera eye can evolve in significantly less than half a million years! This is the blink of an eye, so to speak, in evolutionary timescales, but perhaps this should not have been so surprising. Eyes, it seems, are not the mind-bogglingly complex things we might imagine. They have even evolved, complete with lenses, in single-celled protozoa called dinoflagellates, which hardly possess a complex biological support structure. It seems, then, that the evolution of the eye is a relatively quick and simple process, at least given the underlying chemistry of rhodopsins. It has certainly happened, quite independently, many times since the Cambrian explosion, and there are many creatures living today with radically different solutions to the challenge of vision.

Let us investigate one such animal, admittedly more charismatic than most, whose eyes are similar in sophistication to our own, and yet have a completely separate evolutionary history stretching back well over half a billion years. ◉

# A VERY HUMAN EXPERIENCE OF A VERY HUMAN CREATURE

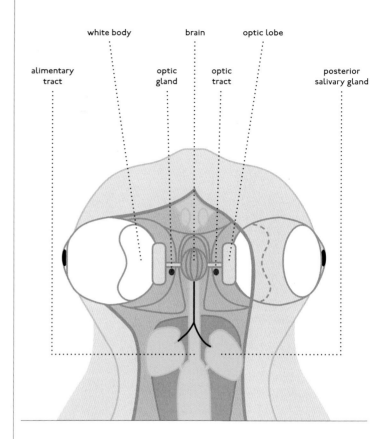

white body · brain · optic lobe

alimentary tract · optic gland · optic tract · posterior salivary gland

We started filming *Wonders* on 22 March 2009 in the arctic cold of northern Norway, and in the three years since then, I've been privileged to visit some of the most beautiful and physically challenging places on Earth. The great lake of lava at Ethiopia's Erte Alle in *Wonders of the Solar System*, the arid high plains of Bolivia in *Wonders of the Universe*, and the ragged, bone-shaking beauty of the journey to Sagada in the northern Philippines in *Wonders of Life* all spring to mind. Such places are chosen because the story drives us to them, and the logistical difficulties are somewhat secondary to the planning process, although they are dragged to the forefront of one's mind when the time comes to roll the cameras. Imagine my delight, then, when we discovered that by far the best animal to bring the story of the evolution of the eye to a close, and to pose one last profound question about the relationship between the visual sense and intelligence, lives in the warm, sandy shallows of Palm Beach, Florida, within sight of a car park beneath the Blue Heron Bridge: Florida's Atlantic shore is an octopus's garden in the middle of a city by a coffee shop.

The octopus is a strange and wonderful thing. Part of the mollusc family, it is a cousin of the slugs and snails in your garden. With no internal skeleton, and an almost entirely soft body apart from its beak, the octopus might appear anatomically bland, but this couldn't be further from the truth. Three hearts pump blue, copper-rich blood around its body, and it is so flexible that it can fit through tiny holes, as long as its eyeballs will pass through – a useful trick when fleeing from a predatory moray eel. As well as ejecting ink clouds to confuse predators, octopuses are among nature's best mimics, able to change their shape and colour to literally disappear into the background. This might appear to make finding an octopus a challenging task, but it is in fact quite simple, because another of the characteristics of the octopus is its curiosity. Aristotle called it 'a stupid creature, for it will approach a man's hand if it be lowered into the water.' We now know that the octopus is far from stupid; its curiosity is borne of intelligence. It is not known how intelligent, or what form this intelligence takes; its brain is the size of a parrot's, but two-thirds of its neurons are not located in its brain, but in its arms. This makes the arms partly autonomous. The brain sends simple signals to the arms, commanding them to perform specific tasks, but the arms themselves contain the 'intelligence' to carry out the details of the tasks on reception of the commands. The octopus has a distributed intelligence totally different to our own, and which evolved completely independently, because our common and almost certainly brainless ancestor predates the Cambrian explosion. The octopus is therefore the closest thing to an alien intelligence on our planet. Despite this, and in common with virtually everyone who encounters one, I found this alien animal compelling, charming and characterful, and the experience of diving with one was, in an indefinable yet tangible way, moving.

As I approached this eight-armed, colourful little being, I must have raised my fists inadvertently, because the octopus did the same. It then scuttled off backwards on six of its arms and, in what I took to be a pretend boxing match, mimicked my movements with the other two. This is well-known behaviour for octopuses; they learn, they mimic, they take a dislike to some people and enjoy the company of others. It's easy to over-anthropomorphise animals, although perhaps less so for something so odd-looking, but it seemed to me that this creature was playful and inquisitive, and possessed of a powerful intelligence.

As an aside, just to illustrate what an impact this little creature had on our film crew, two of our team had octopus tattoos on our first rest day after the film shoot. My director suggested a line for me that I didn't use, because I couldn't say it without donning an embarrassed look not unlike Harrison Ford's during his uncomfortable 'May the force be with you' to Luke Skywalker at the end of *Star Wars Episode IV*. I have nevertheless put his line into practice. 'I will never eat octopus again'.

**LEFT:** Octopuses are well known for their ability to mimic the movements of divers.

**ABOVE:** Two-thirds of an octopus's neurons are located in its arms, rather than its brain.

For our purposes, the octopus is most fascinating because of its eyes. They are in some respects remarkably similar to our own. They are camera eyes, with a lens, iris and retina. There is a reason a camera also has a lens and an iris – the physics of light determines that this is the best way of forming a bright image on a screen, and it is no surprise that evolution by natural selection has been forced towards a common engineering solution more than once. There are, however, notable differences. The octopus doesn't change the shape of its lens as we do to focus on an object. Instead, it works in the same way as an SLR camera, moving the entire lens forwards and backwards to change focus. The retina is also different. While ours is seemingly plugged in back to front, with the photosensitive cells pointing into the eye and the neuronal wiring coming out at the front, the octopus eye has the photoreceptor cells at the front of the retina, pointing directly towards the light. This difference is mirrored in the embryonic development of the eye, which forms from an inward folding of the skin rather than an outgrowth of the brain. Despite these profound structural differences, however, we share the genes responsible for eye development, including the well-known Pax6 gene, also responsible for part of the development of the brain in all vertebrates and invertebrates.

Perhaps the most intriguing of all similarities, however, lies deep within the biochemistry of the photoreceptors – but that similarity also hides an intriguing twist.

**LEFT:** The octopus is a remarkably compelling animal, and the experience of diving to see one can be a strangely moving encounter.

**BELOW LEFT:** The octopus doesn't change the shape of its lens, as we do. Instead, it moves the entire lens backwards and forwards in order to focus.

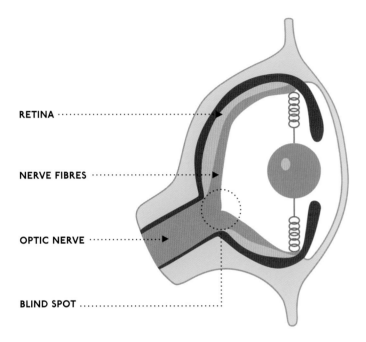

**VERTEBRATE EYE**

RETINA

NERVE FIBRES

OPTIC NERVE

BLIND SPOT

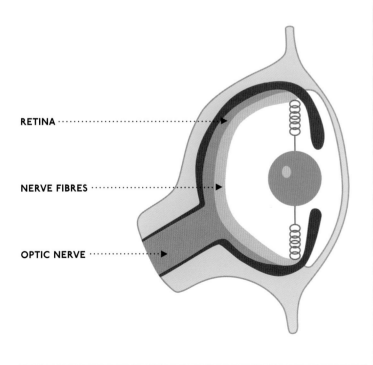

**OCTOPUS EYE**

RETINA

NERVE FIBRES

OPTIC NERVE

213

*The evidence suggests that the common ancestor of all vertebrates and invertebrates possessed both melanopsin and rhodopsin*

The octopus, like all animals, uses a variant of rhodopsin to sense light. But the particular form in the octopus's retina is closely related to the melanopsin we use in the photosensitive retinal ganglion cells that control our circadian rhythms. In fact, all invertebrates use melanopsins for vision. Also, fascinatingly, all invertebrates use rhodopsins for their circadian clocks, while as we have seen, all vertebrates, including humans, do the reverse. This difference manifests itself in the radically different wiring and embryonic development of the octopus eye and our eyes, but the fact that both animals use both types of opsin, not to mention the same controlling genes, is again strongly suggestive of a common origin of vision.

Bringing all this together, the evidence suggests that the common ancestor of all vertebrates and invertebrates possessed both melanopsin and rhodopsin, both originally derived from a common type of rhodopsin that may have been present in cyanobacteria. Perhaps that ancestor used one form for its circadian clock, and the other for the primitive light sensing which predated the evolution of the eye. For some reason, which may have been pure chance, vertebrates like us and invertebrates like the octopus pressed the two forms into use for different purposes. But the smoking gun of a common heritage still persists in the quite staggering similarities across all organisms, in both the biochemistry and the genetic underpinnings of vision. ◉

# COMPOUND VS CAMERA: THE ADVANTAGES AND DISADVANTAGES OF THE SIMPLE AND THE COMPLEX

We have seen that all eyes share common biochemistry, but we have also seen that the overall structure of eyes varies enormously because eyes themselves have evolved many times. Today, among the myriad different forms, there are broadly two types of complex eye, and we've met them both in this chapter; the camera eyes of the octopus and the compound eyes of the mantis shrimp. By the time trilobites died out about 250 million years ago in the so-called Great Dying at the end of the Permian, in which over 95 per cent of life in the oceans became extinct, compound eyes with their multiple lenses had proliferated across a vast array of life. Even today, the compound eye is still the most common form of eye on Earth. Insects, spiders and crustaceans see the world in this way, and although the mineral lenses of the trilobites have long since been discarded, the basic structure of the compound eye has remained the same for hundreds of millions of years. It is a design that has both advantages and disadvantages over camera eyes.

Its main weakness is that the laws of physics limit compound eyes to a relatively low resolution, at least for an eye of manageable size. Each lens behaves like a single pixel, and so the maximum resolution of the image is determined by the number of lenses in the eye. There are only two ways to increase the number of lenses; to squeeze more lenses into each eye, or to make each eye bigger. There is a fundamental limit to the useful size of a lens, however, which is known

*Many insects and crustaceans have compound eyes that entirely dominate their heads. If you wanted to have a compound eye with resolution as good as a human eye, it would have to be 14 m across.*

**BELOW:** The compound eye of a housefly (*Musca domestica*) is good at detecting movement, but does not provide a particularly sharp image.

**BELOW RIGHT:** Scanning electron micrograph of a section through the lens of a human eye, showing the flattened cells arranged in concentric rings.

215

as the diffraction limit. If the lens is too small in relation to the wavelength of visible light, it will not produce a sharp image. The only other solution would be to make the eyes bigger, so that more lenses could fit across its surface, but there is clearly a limit to the size of an eye in relation to the size of the animal itself. Many insects and crustaceans, from the common housefly to the mantis shrimp, have compound eyes that entirely dominate their heads and are pretty much as large as they can be. A compound eye with resolution as good as a human eye would be 14 m across! This is a restriction driven by the physical behaviour of light, and therefore even evolution cannot circumvent it. So resolution is not what compound eyes are about. Why, then, are they so commonplace in nature? One answer is that, once they began to evolve, they couldn't de-evolve and be replaced by something better. Evolution doesn't work like that. Contrary to popular misconception, evolution does not provide optimal solutions to problems; it works with the tools it has, some of which may have arisen by pure chance and are in no sense 'optimal'. This is one of the almost infinite numbers of arguments against the so-called 'intelligent design' fairy story; if a designer were responsible for the structures we see in living things today, he, she or it didn't really think things through properly! Having said that, one has to be careful with making glib assertions about the effectiveness of particular evolutionary solutions, because they often have hidden

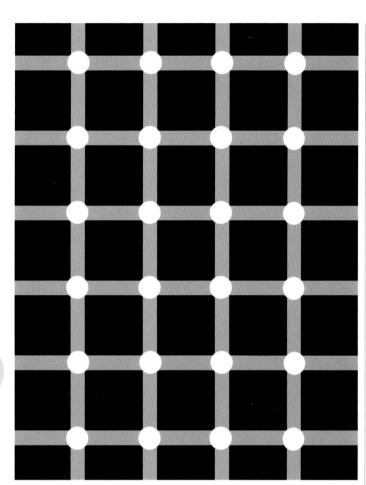

*The advantage conferred on animals that could use their senses to build a detailed picture of their surroundings, and use that picture to hunt, or to avoid being hunted, must have played an important role in the development of intelligence.*

advantages. In the case of compound eyes, the multiple lenses can be 'wired up' so that the eye itself carries out a great deal of visual processing without having to trouble the brain, reducing the demands on the central nervous system when it comes to image processing. This means that insects can react extremely fast to visual stimuli, allowing them to survive and flourish without being clever. A humble housefly, as you may have noticed, can dodge your intelligently controlled attempts to catch it with consummate ease.

So there is a quid pro quo between high-resolution vision, speed of reaction and processing overheads, and in this context it is not so surprising that evolution has arrived at a series of different solutions. Animals such as the octopus and ourselves use a third of our brain power to process the cascade of information delivered by our sophisticated eyes. As well as requiring the grand infrastructure of our complex central nervous systems, this also makes us rather slow to react to visual stimuli compared to the vastly simpler housefly.

One of the more startling examples of what can be achieved by pre-processing visual data within the eye, rather than in the brain, was discovered in a series of experiments on toads in the 1960s. Toads have camera eyes not too dissimilar to our own in structure, but the eyes make fewer demands on their small brains. Toads react very quickly to long, thin objects that move horizontally across their field of view; things that look like worms. But if the same long, thin object is rotated into the vertical plane and moved around, the toad simply doesn't see it. Similarly, if toads are placed in tanks full of dead worms, they will starve because they do not register them as worms. The reason for this apparently eccentric behaviour lies in the way that their photoreceptor cells are wired up. They are connected so that they only fire a signal to the brain when very specific patterns of movement are detected. Long, thin things moving horizontally across the visual field fire a signal to the toad's brain which, in effect, says 'worm'. The toad has little else to do but initiate its hunting response, eating the worm very quickly and efficiently with minimal thinking time. It doesn't have to go to all the slow and complicated trouble of building a picture of the world and searching for a thing that is 5 cm long, pink and cylindrical, within a visual landscape. Its eyes do most of the work, simply by the way the visual cells are connected together. The toad has another system tuned to detect little black blobs that dance intermittently around against a lighter background. This is the fly detection system, and it works with similar speed and efficiency. It is thought that, by focusing purely on a blob that moves like a fly, the toad is able pick out insects against the very complex, rapidly changing backgrounds of its habitat. We would struggle to pick out an insect against forest grasses and trees, illuminated by the shifting light of the Sun and dancing in the wind. But the toad doesn't even see the background. It only sees a black, contrasty blob that moves like an insect, which is all it needs to see to get a meal.

This link between the sense of vision and processing power leads us to the final part of our story of the evolution of the senses. Complex camera eyes produce vast amounts of data that must be processed by the brain. So difficult is this task that, in common with the octopus, around a third of our brain power is used for visual processing. But just how intimate is this connection? Could it be that the evolutionary advantage conferred by increasingly efficient senses was so powerful that costly increases in the sophistication of the data-processing centre, the brain, were also selected for? It's certainly plausible, although the vast leap in intellectual capability between animals such as the octopus and primates was certainly partly due to other selection pressures.

Some scientists believe that foraging for food in harsh and changeable seasonal environments selects for memory and therefore intelligence; that wooded grove was rich in hidden berries or mushrooms last autumn, so we should go back this year. Fine manipulative skills, making it possible

to dig for difficult-to-reach roots and hidden sources of food, may have also played a role. Other scientists believe that more importance should be attributed to the evolution of social groups and cooperation, which requires hierarchies and the development of so-called Machiavellian intelligence, which in turn allows certain individuals to manipulate others and build social structures based on brain power rather than simple physical prowess. But if we look back a long time, before the exponentiation of the intellectual capabilities of our species, it seems clear that the advantage conferred on animals that could use their senses to build a detailed picture of their surroundings, and use that picture to hunt, or to avoid being hunted, must have played an important role in the development of intelligence. ◉

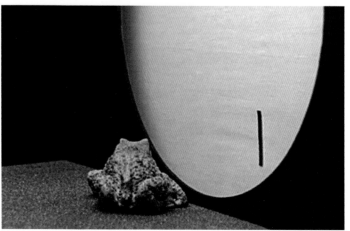

# SEEING THE
# UNIVERSE

T he physicist Freeman Dyson coined the phrase 'Infinite
in all directions' to describe our Universe. We have
attempted to measure the size of quarks, the smallest
known building blocks of matter, and found them to be no
larger than a million million millionths of a metre across. For
all we know, they could be infinitely small. We have measured
the size of the Universe, and found that the part we can see
is 93 billion light years across, and there may be an infinite
universe beyond the horizon that we will never be able to
perceive. As Dyson suggests, we have discovered that we live
out our lives in an intermediate world somewhere between the
infinitesimal and the immense. But most importantly, we have
discovered this; we know it because we used our intelligence,
borne of our senses, to expand those senses with microscopes
the size of cities and telescopes in space. I think there is a
wonderful feedback at work here. As our senses, artificially
extended by our intelligence, deliver more data, we become
more curious about our universe and extend our senses still
further towards infinity in both directions, in a fortunately
futile quest to quench that appetite.

In a way, the great twin disciplines of science and
engineering have taken the place of evolution, allowing us to
rapidly sidestep the biochemical and mechanical constraints
of our bodies and probe worlds that would have remained
forever beyond the grasp of the merely organic. They have
also allowed us to discover how our ability to sense the world
beyond our bodies emerged from the oceans of ancient Earth.
This is one of the great achievements of modern biology;
we now understand that we share the underlying mechanisms
of touch, taste, vision and hearing with every living thing
that possesses these abilities, from the simplest paramecium
to alien underwater animals separated from us by half a
billion years or more. This commonality tells us how the
senses evolved. Far from being a challenge to science, reason
and rational thought, the story of the evolution of the eye in
particular, in all its fine detail, is one of the jewels in the crown
of the scientific method. It connects us, we humans who can
so often feel disconnected from the natural world, directly to
a web spanning all the great kingdoms of life and back, beyond
the Jurassic, the Permian, the Devonian and the Cambrian,
to the first complex cells over 2 billion years ago. And that is
truly wonderful. ◉

**RIGHT:** The US astronomer Edwin
Hubble (1889–1953) worked at
the Mount Wilson Observatory
in California (shown here) for most
of his life. It was while using the
2.5-metre telescope in 1923 that
Hubble observed Cepheid variable
stars in the Andromeda nebula,
proving it was a galaxy outside our
own. Hubble's Law, introduced in
1929, states that galaxies move apart
faster the further away they are –
key evidence for the expansion of
the Universe.

CHAPTER 5
—

# ENDLESS FORMS MOST BEAUTIFUL

## A UNIVERSAL COMMON ANCESTOR

*On the Origin of Species* is without doubt one of the great scientific achievements, the result of the powerful synthesis of careful observation and clarity of thought. Darwin was able to put aside the weight of thousands of years of dogma to reach a revolutionary conclusion: 'Therefore I should infer from analogy that probably all organic beings which have ever lived on this earth have descended from some one primordial form, into which life was first breathed.' To the nineteenth century mind, this must have been a shocking thesis. Humanity did not arise intact from the dust; we were not created intact. Instead, our lineage, in common with every other living thing on Earth, can be traced back to some long-forgotten population of simpler organisms. We are related, literally, to every animal, plant and bacterium cell alive on the planet today.

**RIGHT:** Charles Robert Darwin
(1809–1882) is most famous for
his theory of evolution, published
in 1859 in *On the Origin of Species.*

There are few things more satisfying in science than the existence of an iconic number. The kind of number that requires a deep understanding to uncover, and expresses something profound about the properties or structure of the Universe. Such numbers occasionally come with units and uncertainty attached: $13.75 \pm 0.11$ billion years is one such number – our best current measurement of the age of the Universe. The number itself would be meaningless to a visiting alien, because a year is an arbitrary and continuously variable unit of time based on the details of our wobbling orbit around the Sun. But it is still an iconic number, because to know it requires precision observations of the recession velocities of countless distant galaxies, an understanding and measurement of the cosmic microwave background radiation, the light released 380,000 years after the Big Bang, and Einstein's general theory of relativity, one of the greatest achievements of twentieth-century science. There are, in other words, centuries of engineering achievement and theoretical and experimental understanding wrapped up inside this single number.

There are also numbers that are built into the very fabric of our Universe – mathematical constants such as: $\pi$, 3.14159..., the ratio of a circle's circumference to its diameter, or the fine structure constant $\alpha$, approximately equal to 1/137, which encodes the strength of the force of electromagnetism.

In biology, there is a number of fundamental interest that has proved extremely difficult to measure; even its order of magnitude is a matter of genuine scientific debate. Ask a biologist how many species live on planet Earth today and the chances are you will get a shake of the head because we simply do not know how many species share our home. A recent estimate put the number at 8.7 million, but the literature is replete with criticisms of the methodology, illustrating how far away we are from a consensus view. Other estimates range from 3 million to 100 million extant species. What is known is that 1.3 million species have been catalogued, of which we are one, and that number is climbing at a rate of approximately 15,000 per year. ◉

# EVOLUTION AND MADAGASCAR

**BELOW:** Madagascar's lemurs provide an example of the origin of species.

**RIGHT:** The sheer quantity and diversity of life found in Madagascar reflects its range of habitat and climate.

There are places that lodge in the imagination. They settle into a niche in the mind that allows them to grow and branch, catalysing new ideas and precipitating a revisionist understanding of the original experience that runs far deeper than the immediate sensory impression. Such experiences are rare, created only by precious places. For me – I insert the caveat because these choices are subjective and personal – Madagascar was such a place. This island the size of France, separated from East Africa by the Mozambique Channel, is home to over a quarter of a million species, of which 90 per cent are found nowhere else. This diversity reflects Madagascar's range of terrain and climate. The southeastern trade winds keep the eastern coast wet, particularly between November and April, and deliver powerful and destructive storms. The central highlands protect the western coasts, leading to semi-desert conditions in the southwest. The capital Antananarivo, known to the locals and grateful, tongue-tied tourists as Tana, sits in the highlands astride these zones, experiencing a relatively wet rainy season and dry, chilly mornings from May to October. Tana in June is a city of hills, valleys and deep colours made vivid by the lambent winter sun. Rice paddies punctuate and disrupt the outskirt sprawl, lending the capital an unusually spacious feel. The elevated central areas are French colonial, the old buildings always appropriately framed by a museum of rattling pale cream Citroën taxis.

One should resist the temptation to over-glamorize, of course. The ancient island is a poor country, with 90 per cent of the population living on less than 2 dollars per day. It is also a country that is actively threatening its own riches. Around 1 per cent of Madagascar's forests are destroyed every year, primarily through the practice of 'Tavy', slash and burn agriculture aimed at clearing land for rice farming and charcoal production. During the second half of the twentieth century, over half of the forests have fallen, lending large swathes of the country a bleached and barren appearance from the air. In the context of Madagascar's unique ecosystem, this is a tragedy of global proportions, caused by poverty undoubtedly, but also a fragile political system battered, I was informed by enough Malagasy to be confident in reporting it, if not as fact, then at least as a valid and widely-held opinion, by continual post-colonial meddling.

I am aware that the reader may have read many expressions of regret at the vanishing of habitats and species across the globe, and that familiarity with an idea can breed contempt. But for me, the experience of making this film made it personal, not just because of people and place, but also because Madagascar provides both an impressionistic backdrop and literal example of a series of extremely important ideas. These ideas are absolutely central to a lesson – yes, lesson is the right word, carrying as it does a hint of wagging finger – which our scientific exploration of the world has delivered. The lesson is this; the tree of life on Earth is unique, and therefore incalculably valuable. The ideas underlying this lesson are those first expressed in Darwin's *On the Origin of Species*. Together with our modern understanding of biochemistry and genetics, these ideas explain the existence of our unique tree. And it is through an understanding of how our tree emerged that a true appreciation of the value of every branch and twig emerges.

Let us begin our exploration of these grand ideas in the forests of eastern Madagascar. ◉

226

# DARWIN'S BARK SPIDER

Madagascar is home to over a quarter of a million species, 90 per cent of which are found nowhere else. We chose one of the recently catalogued species to begin our film. *Caerostris darwini*, to give its official name, was identified as a new species in 2009 by a team of zoologists from the University of Puerto Rico exploring the mountain forests of Eastern Madagascar. The name was chosen to celebrate the 150th anniversary of the publication of Darwin's *On the Origin of Species* on 24 November 1859. The spider, more commonly known as Darwin's bark spider, exploits a particular environmental niche: it hunts where no other spider can, by taking the art of spinning webs to a new level. Its webs are constructed to the same orb design as those of a common garden spider, but the size is quite different; Darwin's bark spider spins the largest-known webs on the planet. The main anchor lines can stretch up to 25 m across mountain rivers and streams, and, suspended in the centre, the central core of a web can cover a diameter of 2 sq m. It is tempting to imagine that this allows the spider to catch huge prey, like birds or bats, but this fear, which plays on the mind of the uninformed traveller in the mountains of Madagascar, is immediately dispelled by a fleeting glimpse of the engineer. The spider itself is tiny in comparison to its home; the females are only around 2–3 cm in length, while the males are much smaller.

The advantage these spiders possess, therefore, is not their physical prowess, but the strength of their silk. It is the toughest-known biological material; twice as elastic as that of any other spider. When stress-tested to breaking point in the laboratory, it was found that the threads are ten times tougher than Kevlar, the material used for bullet-proof vests. This adaptation confers an advantage on these little spiders, because it allows them to farm the airspace above rivers – rich in dragonflies and mayflies, and devoid of competition from other spiders. In other words, they exploit a niche in the otherwise inaccessible landscape above the water.

In the final paragraph of *On the Origin of Species*, Darwin finds a vivid metaphor for the natural world in an entangled bank, 'clothed with many plants of many kinds, with birds singing on the bushes, with various insects flitting about, and with worms crawling through the damp earth…'. It is interesting to reflect, he writes, 'that these elaborately constructed forms, so different from each other, and dependent on each other in so complex a manner, have all been produced by laws acting around us.' His central idea is beautifully illustrated by *Caerostris darwini*. Let us imagine that a spider is born that produces slightly stronger silk, and that this ability, once it appears, can be passed down from generation to generation. Darwin didn't know anything about the underlying mechanisms, which we now know are made possible by the unique properties of DNA. But the detail doesn't matter for Darwin's argument. What matters is this: if an inherited trait confers some advantage in the 'struggle for life', then that trait will proliferate in future generations, simply because it is more likely to be passed on to those generations. This leads, in Darwin's language, to a 'Divergence of Character and the Extinction of less-improved forms.' This, in a nutshell, is natural selection. Stronger silk opens up a new niche, which allows access to more food, and makes it more likely that the possessor of stronger silk will survive long enough to transfer this trait to its offspring. The animals' interaction with the environment and other animals acts like a sieve, which selectively tests new traits and preferentially allows those that confer some advantage to pass, purely by dint of statistics. In the intensely competitive world of Darwin's tangled bank, any trait that increases the chances of producing offspring and can be transferred to those offspring will ultimately result in the predominance of that trait in the population. Technically, and this is an important point well illustrated by Darwin's bark spider, selection occurs on the phenotype, which is the term for the organism's physical properties and its behaviour and constructions. It is the web, as well as the spider, that is passing through the sieve, although the instructions to produce the silk are ultimately contained within the spider's genetic code.

This, in a nutshell, is the subject of this chapter. But it is admittedly quite a dense series of ideas. So let us unpack Darwin's great insight, using the animals of Madagascar and neighbouring South Africa as our guide. ◉

*Darwin's bark spider spins the largest-known webs on the planet. The main anchor lines can stretch up to 25 m across mountain rivers and streams, and, suspended in the centre, the central core of a web can cover a diameter of 2 sq m.*

**LEFT:** The web of *Caerostris darwini* (Darwin's bark spider) spans the upland rivers and streams of Madagascar, and its silk threads are ten times stronger than Kevlar.

# DARWIN'S ORCHID

One of the best illustrations of the power of selection, which also demonstrates the speed with which the form of an organism can change, is not natural but artificial. Think of the example of the breeding of the domestic dog. Humans began selectively breeding dogs from wolves around 15,000 years ago. By selecting certain individuals for breeding, based on traits such as short legs, long fur or powerful jaws, and explicitly preventing random interbreeding between the offspring, today's quite startling array of forms, from Poodles to German Shepherds, has been produced. If there had been no artificial selection and separation by humans, we would still be left with wolves. The word 'separation' in the previous sentence is very important in the context of the evolution of new species in nature. For the example of domestic dog breeds, it should be obvious why; interbreeding washes out differences between breeds, and can lead to a large degree of homogenisation. We will return to this point later on, but suffice to say that our choice to film on the island of Madagascar is not only down to its unique animals, but also because it is a physical island that separates populations from mainland Africa in much the same way as humans separate different breeds of dog.

For now, the key point is that artificial selection produced a quite dazzling array of different dogs, many fine-tuned to specific tasks such as hunting or herding sheep. And it did this very quickly indeed, over a time period of a few thousand years, with a great deal of change occurring in the last two hundred years or so.

In nature, precisely the same processes occur. Natural selection replaces artificial selection, of course, but there is no difference in principle. The point is that a particular characteristic or adaptation that makes it more likely that an organism with this adaptation will survive, will eventually make it more likely that organisms with this adaptation will breed together, simply because there are more of them. The result will be that the characteristic will tend to predominate in the population. The natural world can perform this task just as effectively as humans. To make this point crystal clear, it is worth considering that the rich variety and beauty of flowers is partly the result of selection by insects and birds, and the opposite is also true: flowers apply selection pressure on insects. One of the most famous examples can be found in the lowland forests of Madagascar.

The *Angraecum sesquipedale* is a flower with a special, Darwinian history. First discovered by the French botanist Louis-Marie Aubert du Petit-Thouars in 1798, this orchid grows attached to the trunks and branches of trees and produces beautifully structured white flowers once a year. Although in its native southern hemisphere habitat it flowers between June and September, when cultivated in Europe the flowers appear between December and January, earning it the name the Christmas orchid. Darwin studied many species of orchid and wrote about them extensively, publishing an entire book on the subject in 1862 with the title: *On the various contrivances by which British and foreign orchids are fertilised by insects, and the good effects of intercrossing.*

Darwin was fascinated with the intimate connection between the form of insects and the pollination of flowers such as the Christmas orchid, and this drove him to make one of his most famous predictions. The spur of the Christmas orchid is extraordinarily long; it can be up to 40 cm from the tip of the flower to the nectar contained deep within. Darwin knew that the purpose of nectar is to attract the attention of insects, which dutifully transfer pollen from one flower to another. This implies that an insect must exist with a proboscis long enough to reach deep inside the flower. When seeing the orchid for the first time, Darwin is reported to have responded with the immortal words 'Good heavens, what

**ABOVE:** The extra-long proboscis of *Xanthopan morgani praedicta* (Morgan's sphinx) reaches deep inside Darwin's orchid, demonstrating the intimate connection between the form of the insect and the pollination of the plant.

insect can suck it?' In his wonderfully titled book of 1862, Darwin duly predicted the existence of a moth with a 40 cm long proboscis. Twenty-one years after his death, and forty years after his prediction, a moth fitting Darwin's description was found in Madagascar by Walter Rothschild and Karl Jordan, who in fact cited a later and more precise prediction by Alfred Russel Wallace, namely that the moth should be a hawk moth similar to those found in East Africa. Despite Darwin's prescience, the moth was duly named *Xanthopan morgani praedicta* – with the '*praedicta*' referring specifically to Wallace's prediction. Darwin is honoured in the other common name of the orchid – Darwin's orchid. ◉

# A NAME FOR LIFE

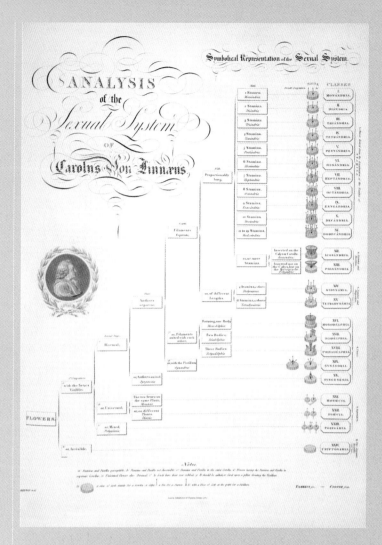

Whenever a new species is discovered and named it joins a system of classification that has been in existence for almost 300 years, a system that began with the work of one man Carl Linnaeus. Linnaeus is the father of modern taxonomy, designing the system of classifying, categorising and naming life that is still used to this day. In fact, when it comes to taxonomy there is one simple divide – before Linnaeus and after him. Born in the Swedish village of Råshult in 1707, the young Linnaeus arrived into a family that was obsessed with the two things that would later define his career and legacy. In the Linnaeus household names weren't just important – they were revolutionary. At this time it was common in Scandinavian countries to follow a patronymic naming system – your surname was your father's first name and changed with every generation. Centuries of tradition were about to be broken, however, when Carl's father Nils gained a place at the University of Lund and decided to adopt a permanent family name. As a keen amateur botanist, a passion that he would later pass to his eldest son, Nils didn't have to look far for inspiration. A giant lime tree (Latin name 'Linnaeus') sat proudly on the family's land and so this was the name he chose. Carl, the first born of the family, was the proud recipient of this newly minted title, a name that lives on in the 'Linnaean system' of taxonomy that we use to this very day, and it is fitting that it should have such a biological origin.

The young Linnaeus soon became enthused with his father's passion for the natural world, working with him in the garden and discussing the names and qualities of the many flowers that grew there. He was quickly given a piece of earth to grow his own plants, and so began a relationship that would transform our understanding of the natural world and bring order to the seemingly chaotic variety of life. It is remarkable to think that this is a man who has ultimately left his mark on every living thing we have discovered on the planet, and yet, to begin with, this rulebook was strewn with errors.

**CLASSIFICATION SYSTEM DEVISED BY CARL LINNAEUS**

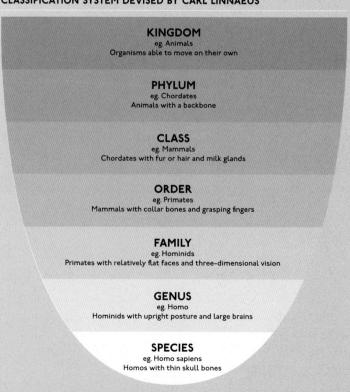

**KINGDOM**
eg. Animals
Organisms able to move on their own

**PHYLUM**
eg. Chordates
Animals with a backbone

**CLASS**
eg. Mammals
Chordates with fur or hair and milk glands

**ORDER**
eg. Primates
Mammals with collar bones and grasping fingers

**FAMILY**
eg. Hominids
Primates with relatively flat faces and three-dimensional vision

**GENUS**
eg. Homo
Hominids with upright posture and large brains

**SPECIES**
eg. Homo sapiens
Homos with thin skull bones

231

The first edition of his greatest work, *System Naturae*, was published in 1735, but it took ten editions and over twenty years of work before its position as the bible of nomenclature was reached. In the meantime, Linnaeus continued to name and categorise not just newly discovered species but also the mythical creatures of the time, which held such powerful sway in the mind. The phoenix and dragon featured in early editions, as well as the sphinx-like manticore, with each of these fabled creatures being gathered under the collective grouping of 'Paradoxa'. It gave Linnaeus's initial work an air of the medieval bestiaries that were so popular at the time, and it wasn't until the 6th edition, published in 1748, that Paradoxa was entirely removed from his classification system. In many ways it appears that Linnaeus was actually trying to debunk these creatures by including them in his works, a way of removing the mist that shrouded a world filled with magic and superstition. Classifying humans was another contentious part of Linnaeus's early work, with humans being organised in the same grouping as other primates, a bold statement over one hundred years before Darwin published his heretical work. Our genus *Homo* also included another species, *Homo troglodytes*, a caveman-like creature, and *Homo lar*, a human-like form covered in fur that is today known as the lar gibbon. Despite these many errors, over a period of 40 years Linnaeus laid down the foundations of the system of kingdoms, phyla, classes, orders, families, genera and species (see illustration opposite) that biologists employ around the world today. Pivotal to this system is the existence of a lectotype, a single specimen that can be used to identify and represent the critical characteristics that define a whole species. Linnaeus collected many lectotypes in his career and published details of thousands of them in his books, and it was this system that has perhaps provided him with the greatest honour of all. It wasn't until almost 200 years after his death, but since 1959 Carl Linnaeus has been the official lectotype of *Homo sapiens*, the representative specimen of all humans on the planet.

At the time Linneaus was honing his system of classification, about 10,000 species had been identified by science, with roughly 6,000 plant species and 4,000 animal species contained within the literature. Over the last 250 years that number has climbed to 1.3 million characterised species, and we still don't know exactly where this number will end or even if it is possible to actually classify every life form on the planet. Even so, this doesn't stop science from adding around 15,000 new species to our collective knowledge each year – a job that is becoming all the more important as the habitats that hide so many undiscovered species are destroyed across the world. ◉

ENDLESS FORMS MOST BEAUTIFUL

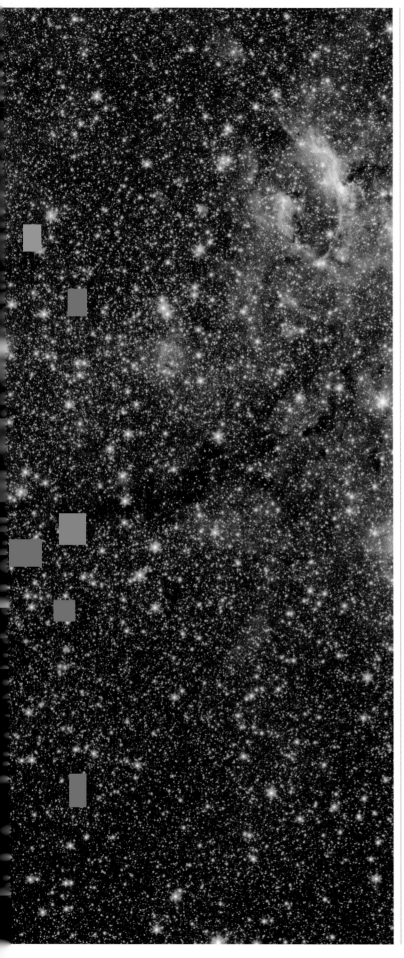

# THE STUFF OF LIFE

Few elements on earth are as abundant, useful, or indeed as vital as carbon. It is an element that occurs naturally in many forms; from charcoal and graphite to rare and precious diamonds. Carbon fuelled the industrial revolution, transformed the architecture of our cities by turning iron into steel and redefined what is possible in engineering with the invention of carbon fibre. Today we find ourselves on the verge of another carbon-powered technological revolution with the discovery of graphene, a one atom thick sheet of carbon atoms with extraordinary characteristics. But carbon is also, of course, the element of life.

*Every living thing is constructed from molecules containing carbon. It is the scaffolding of life.*

Every living thing is constructed from molecules containing carbon. It is the scaffolding of life. Everywhere in the natural world, intricate collections of carbon-based molecules are breathing, breeding, flying, running and thriving. Every protein, carbohydrate and fat molecule in your body is constructed around carbon; from the 100 billion neurones in your brain, to the muscle behind every beat of your heart, to the architecture of DNA, you are rightly described as 'carbon-based'. Almost 20 per cent of your body, by mass, is carbon.

Limitlessly adaptable, every cell of your body is filled with organic compounds and structures based on carbon chemistry, and these are just a fraction of the ten million known organic compounds. From Darwin's bark spider, to an orchid, to an English academic from Oldham, carbon is a necessary building block of life on Earth. ◉

**LEFT:** Through the force of gravity, hydrogen and helium coalesce into enormous clouds that eventually become the balls of gas that form stars.

# THE BIRD: COLLISION OF A TRILLION SUNS

One of the most exciting nights for me during the filming of *Wonders of Life* was a visit to the newly built South African Large Telescope (SALT), in the Karoo region of South Africa. SALT is an engineering marvel; it is the largest optical telescope in the southern hemisphere, using 91 identical hexagonal mirrors operating together to create a reflecting surface of over 66 m². Under crystal-clear African skies, SALT peered 650 million light years away from Earth to take iconic images of 'The Bird', an almost organic-looking object that is the aftermath of the collision of three galaxies.

The wings and tail of the 'bird' were created around 200 million years ago in the collision of two giant spiral galaxies, and span a distance of over 100,000 light years. That is roughly the size of our Milky Way, which contains almost half a trillion stars. A third irregular galaxy is currently careering into the aftermath of this galactic collision at a colossal 400 km/s, forming the head. The shock wave from the collision is causing clouds of interstellar dust and gas to collapse, forming new stars at the rate of 200 solar masses every year. These new stars, and the discs of dust that surround them, will be rich in carbon and oxygen. In time, these discs will condense into planets, and those newly minted heavy elements will become part of rocks, maybe oceans, and – who knows? – life. If that sounds unduly speculative and romantic, so be it, but we know it happened at least once in the Universe. ◉

**BELOW:** 'The Bird', formed by the collision of two giant spiral galaxies, with a third irregular galaxy approaching from the left at a colossal 400 km/s.

**BOTTOM:** The South African Large Telescope (SALT) is located in the Karoo region of South Africa, an area blessed with wonderfully clear night skies.

**RIGHT:** SALT is the largest optical telescope in the southern hemisphere, with 91 hexagonal mirrors that form a reflecting surface of more than 66 sq m.

234

235

# THE HOYLE RESONANCE: ARE WE LUCKY TO BE HERE AT ALL?

The Universe began at the Big Bang, 13.75 billion years ago. Or maybe it didn't. What is known is that something interesting happened 13.75 billion years ago, which resulted in our observable Universe being placed into a very hot, very dense, and extremely highly ordered state which has been expanding, cooling, and getting more disordered ever since. The most widely accepted cosmological theories do indeed suggest that time began at this point, known as the Big Bang Singularity, but there are other models that suggest an eternal Universe, within which an event happened that caused the Big Bang. For the purposes of our story, these details do not matter. What matters is that our Universe used to be hot and

**BELOW:** Nebulae are interstellar clouds of dust, hydrogen, helium and other ionised gases. The Cat's Eye Nebula, shown here, is in the constellation of Draco. Its central star is over 10,000 times brighter than our Sun.

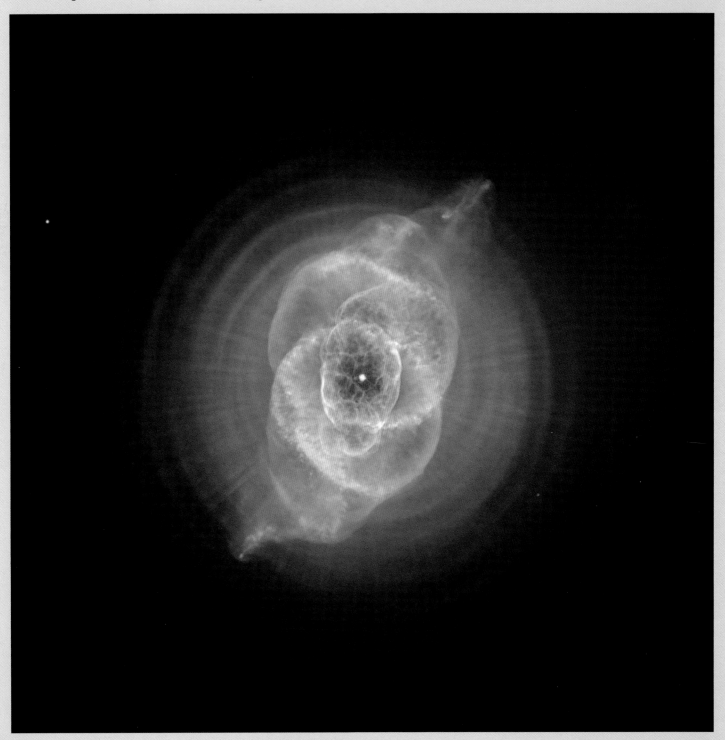

dense, with no galaxies, no stars, no planets, no chemical elements and no subatomic particles at all. Around one second after the Big Bang, the Universe had cooled to a relatively chilly 10 billion degrees – cool enough for the building blocks of the chemical elements, protons and neutrons, to form. After around three minutes, the three lightest elements, hydrogen, helium and lithium, were present, and that is the way the Universe stayed for the next half a billion years or so. All the heavier elements, including carbon, had to wait until the first stars formed, because they were forged by nuclear fusion reactions inside stellar cores.

The basic physics of the synthesis of heavy elements by nuclear fusion is quite simple. Stars begin their lives fusing hydrogen into helium. Hydrogen nuclei consist of single protons. Helium consists of two protons and two neutrons. The first step in the formation of helium is for two hydrogen nuclei to approach each other at high speed, which happens because the temperature in the stellar core is so high – of the order of 15 million degrees Celsius at the centre of our Sun. During this process, one of the protons can turn into a neutron via the action of one of the four fundamental forces of nature known as the weak nuclear force. The proton and neutron then 'fuse' together, which means they are bound together as a pair by the strong nuclear force, in much the same way that an electron is held in orbit around an atomic nucleus by the electromagnetic force. This bound proton-neutron pair is known as deuterium. Very quickly, another proton fuses with the deuterium nucleus to form helium-3, and then two helium-3 nuclei fuse to form helium-4, made of two protons and two neutrons, with the emission of the two 'unused' protons. The key point, from the perspective of the physics of stars, is that the single helium-4 nucleus produced by nuclear fusion is less massive than the four protons that started the process. This difference in mass is accounted for as energy, which is why stars shine. Our Sun will continue to burn in this way for another 5 billion years or so, until it runs out of hydrogen fuel in its core. Then, robbed of the energy released by hydrogen fusion, it will begin to collapse under its own gravity and the temperature in the core will rise, initiating helium fusion. This is the process that creates carbon, but with a fascinating twist.

Carbon production begins with two helium-4 nuclei fusing into beryllium-8, a nucleus with four protons and four neutrons. Beryllium-8 is unstable, and decays back into two helium nuclei very quickly. If another helium nucleus approaches during its short lifetime, however, it has the briefest of opportunities to fuse with the unstable beryllium to form carbon-12. This is the origin of the carbon in your body, and in every living thing on Earth. However, there is a problem with this simple description, first pointed out by the British astrophysicist Fred Hoyle. The rate of production of carbon-12 is, at first sight, vanishingly small, because carbon-12 is slightly too light to 'encourage' a beryllium nucleus to fuse with a helium nucleus. This is rather loose language, of course. More precisely, the rate of nuclear fusion reactions can be greatly enhanced through a process known as resonant production. If the masses of the incoming nuclei are very close to the mass of the produced nucleus, then the reaction will proceed much more quickly.

Hoyle's solution was to propose the existence of what is known as an excited state of carbon-12. This excited state, written 12C*, is built out of precisely the same constituents – six protons and six neutrons – as the more usual form, 12C, but the constituent protons and neutrons are arranged differently, resulting in a slightly higher mass. This means that it is much more likely for a short-lived beryllium-8 nucleus to fuse with another helium-4 nucleus to form the excited state of carbon, which then quickly decays back into standard 12C. Shortly after Hoyle made the prediction in 1953, the 12C* state was identified. Without it, the production rate of carbon in stars would be 10 million times smaller, and there would be very little carbon in the Universe.

There is yet another interesting twist to this story. Hoyle also noticed that a further fusion reaction, adding another helium nucleus to carbon to form oxygen, should also occur, and if this were also resonant, then all the carbon would be consumed as fast as it could be produced. Fortunately, the combined masses of the 12C and 4He nuclei are very slightly above those required to produce 16O resonantly, so the newly minted carbon survives in large quantities.

To appreciate the meaning of this little detour into astrophysics, note that the precise details of the masses of these resonances depend on the strengths of the fundamental forces of nature. Slight changes in the strengths of these forces, in particular shifts of the order of 1 part in 105 in the fine structure constant, which sets the strength of the electromagnetic force, change the rates of carbon and oxygen production in stars significantly. This is known as a fine-tuning problem; if things were even slightly different at a fundamental level, life may well not exist because there would be radically different amounts of carbon and oxygen present in the Universe. Such calculations are notoriously difficult, and this is still an active area of scientific research and debate, but from a historical perspective it is interesting that a property of the carbon nucleus itself was predicted based on the observation that there is a large amount of carbon in the Universe. This is why Hoyle's prediction is often referred to as an 'anthropic prediction'. We exist, and we are made of carbon, therefore the excited state 12C* exists.

Whether or not our Universe is fine-tuned for life, and whether or not we are extremely lucky or just slightly lucky to be here, it is quite wonderful, I think, to appreciate that every single carbon atom in every amino acid, protein and strand of DNA in your body was forged in the heart of a long-dead star, returned to the Universe in either a supernova explosion or through the artistry of a planetary nebula, only to be caught up in a collapsing dust cloud 4.6 billion years ago around a newly formed star we now call the Sun. And we can see that process of star formation from interstellar debris continuing to this day. ◉

237

# CARBON CYCLE

Every single atom of carbon in your body will have been constructed in the heart of a star. It will have seen its parent star die, and escaped from its gravitational pull in the quiescent coloured drift of a planetary nebula or the instantaneous ferocity of a supernova. It will have seen our Sun and Earth form and most probably settled into Earth's rocky mantle for billions of years before entering the atmosphere in the outgassing of some primordial volcano. It will have spent time in the cycle of life before; it is tempting to dream that it spent time in the claw of a *Tyrannosaurus rex* – who knows? – in the way that the victims of charlatanistic past-life regression therapists always seem to discover that they were once Roman legionaries or Cornish pirates, rather than an agricultural peasant who died of cholera. But what we do know is that the carbon atom entered your body via a plant or algae, which you ate either directly, or second-hand as meat.

# THE GREATEST CYCLES OF LIFE

Virtually all carbon enters the food chain today through photosynthesis, the process by which plants and algae construct simple sugars from carbon dioxide and water, using sunlight as a power source.

$$6CO_2 + 6H_2O + photons \longrightarrow C_6H_{12}O_6 + 6O_2$$

This means that each of our carbon atoms, with a very high probability, spent some time floating around in Earth's atmosphere as a molecule of carbon dioxide. Inside a chloroplast, those ancient bacterial machines, it had its stable existence disrupted by the forceful form of chlorophyll that, encouraged by sunlight, forced extra electrons onto the carbon dioxide molecule, opening up those strong carbon-oxygen double bonds and beginning the process of the formation of long carbon chains.

The trees of the forests, plants of the prairies and algae of the oceans therefore sit at the very base of the planetary food web, beginning the process of long-chain carbon molecule formation by doing the difficult job of turning gaseous $CO_2$ into sugar.

Most of us have a strong conscious affinity for the sweet stuff, but all of us need it, because sugar, or D-glucose to give it its correct chemical name, is life's basic energy supply: six carbon atoms in a chain, combined with six oxygen and twelve hydrogen atoms, which would, from an energetic perspective, be much happier as carbon dioxide and water. Life liberates this stored energy and uses it to form ATP through a series of fiendishly complex biochemical pathways. In eukaryotes, this involves the delicate machinery of mitochondria and proton cascades that we met in Chapter 2. Ultimately, a single glucose

**A SNAPSHOT OF A SMALL PART OF A LIGNIN MACROMOLECULE**

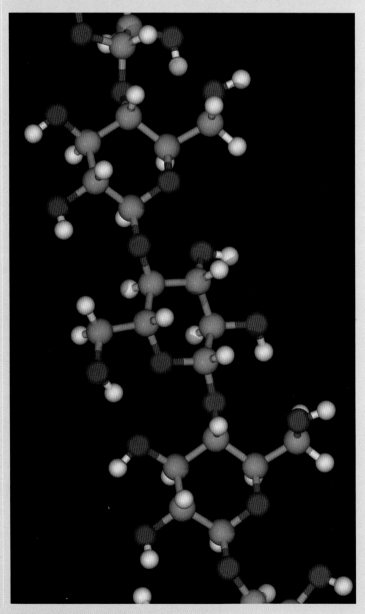

**RIGHT:** Cellulose is a polysaccharide with a very long chain of D-glucose molecules consisting of atoms of carbon (shown in grey), oxygen (red) and hydrogen (white).

**FAR RIGHT:** This scanning electron micrograph of the cut edge of a leaf shows the thick walls of cellulose surrounding each cell. Cellulose is both strong and stable, and is the most common organic compound on Earth.

240

molecule, fully oxidised, typically produces around 30 molecules of ATP – a rich bounty.

In an almost grotesque simplification, then, the oxidation of glucose, which provides the energy to power all animals, looks precisely like photosynthesis in reverse:

$$C_6H_{12}O_6 + 6O_2 \longrightarrow 6CO_2 + 6H_2O + energy$$

It appears that life simply plucks carbon out of the air, transforming it through the alchemy of photosynthesis into the universal food available for all and sundry to feast on. But things are far more interesting than that, because although plants and algae are busy making D-glucose from carbon dioxide, water and sunlight, very little of it stays in this immediately useful form for long. Instead, plants use the sugars to build much longer-chain molecules, including the structural molecules cellulose and lignin.

Cellulose is a long chain of D-glucose molecules known as a polysaccharide, with the general chemical formula $(C_6H_{12}O_5)n$. The n means that a very large number of these units are linked together. Here is carbon functioning as scaffolding, bonding to other carbon atoms over and over again to form a giant organic molecule. Cellulose chains can be over 10,000 units long.

Cellulose is the most common organic compound on Earth, forming the structure of the cell walls in green plants. It is strong and stable, and therefore difficult to break down.

Lignin is even tougher than cellulose. Rather than being composed of long chains, it is cross-linked into complex lattice structures comprising many tens of thousands of atoms. Wood with high lignin content is extremely resistant to degradation by biological action, which is why boats and houses are made of it.

Most of the carbon and energy stored away by plants is therefore immediately converted into a form which life finds difficult to access. Most mammals, including humans, have limited ability to break down cellulose, and there are no animal enzymes that can digest lignin. So it would seem that the entrance of carbon into the food chain via photosynthesis is rather more convoluted than might have been expected at first sight. ◉

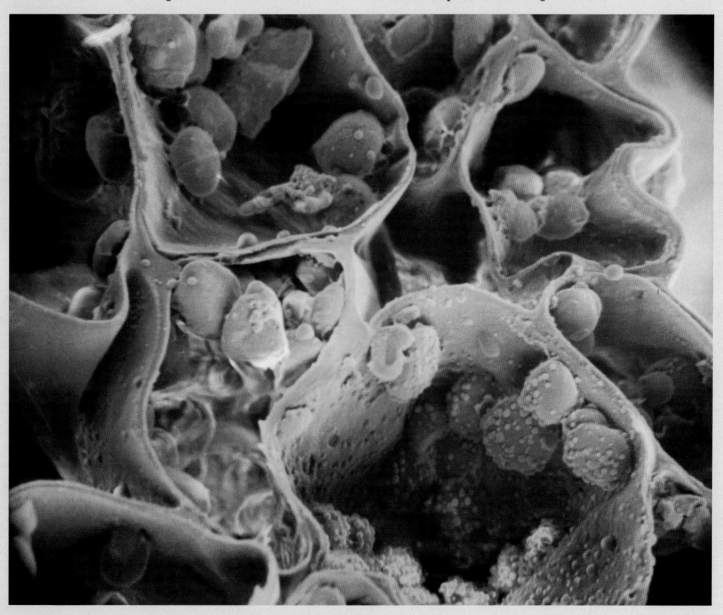

# HARVESTING CARBON

The Kruger National Park is the Africa of the imagination. Occupying almost 20,000 sq km in northern South Africa, bordering Mozambique and Zimbabwe, here zebra, giraffe and wildebeest are as common as sheep are on the hills of northern England. The short winter days give way to evenings of reddening light and volume that rises with the setting Sun, a mounting siren choir of insects ferociously punctured by noises native to the gathering African night. The Dark Continent is seductive, but be in no doubt that nature at this intensity is no longer a place for the human species; few refugees from Mozambique who follow the power lines across the Eastern fence make it across the veld to Nelspruit and onwards to Pretoria and Johannesburg. Kruger is a fence around a frightening past, but from a Land Rover driven by a man with a gun, it is beautiful.

Kruger's latitude is 24 degrees south, and its landscape is shaped by the turning of the seasons. Many of the trees are deciduous, and in the winter months the fallen leaves and dry grasses mean that the supply of food for plant eaters is at low ebb. The easily accessible sugars of the green plants are long gone, leaving the tough lignin and cellulose that make up the woody stems and roots of a parched terrain.

At first sight it might appear that this seasonal change builds barriers to the flow of carbon from atmosphere to plant to animal, but plant eaters come in all shapes and sizes, and some of the smallest have evolved the most remarkable strategies to enable them to attack the tough remnants of photosynthetic growth.

The sculptured mounds of termites have been a characteristic feature of the African landscape for over

**ABOVE:** Termite mounds, which have been a characteristic feature of the African landscape for over 200 million years, are the only visible part of the termite colony that spreads underneath it.

**RIGHT:** Complex symbiotic relationships can exist between different species: here *Termitomyces* (a genus of fungus) breaks down the termites' faecal pellets, which contain undigested lignin, into a form that the termites can then digest.

## TERMITE MOUND CROSS-SECTION

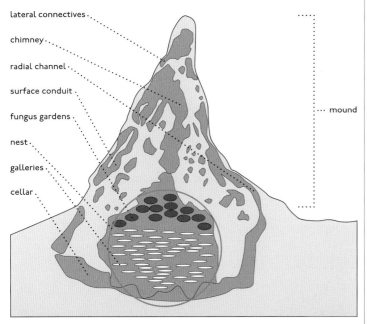

lateral connectives
chimney
radial channel
surface conduit
fungus gardens
nest
galleries
cellar

mound

200 million years. Closely related to the cockroach family, termites are eusocial insects. Like ants and bees, they live in vast colonies of millions of individuals. Workers, soldiers and queens all have specific roles within the highly organised society. But this is not a directed union, at least in the sense that our behaviour is directed from our brain. The termite colony exhibits complex behaviour that emerges from simple rules governing the interactions between individuals and the outside world. I suspect the previous sentence comes as no surprise to most readers, and might pass unnoticed. But the intricacy and complexity of the colony and the precision of the engineered mounds is, to my mind, more deserving of the label 'Wonder of Life' than the spectacular and more photogenic predators of the Kruger.

The mounds are only the visible part of the termite colony, which extends well below ground. It functions as an air-conditioning system, channelling the wind through tunnels, which can be opened or closed to keep the internal temperature of the mounds constant to within half a degree. Humidity is also carefully controlled. It is thought that the termites dig wells deep into the earth to access ground water, which they carry upwards into the chambers of the colony. So effective are these high-precision, passive air-conditioning systems that there are ongoing research projects aimed at transferring termite designs to large buildings in human cities.

**ABOVE:** Termites live in vast colonies of millions of individuals, and workers, soldiers and queens all have specific roles within the highly organised society.

The reason for the precision climate control is perhaps even more fascinating than the mechanism. These termites need to create and maintain the conditions of a rainforest in the dry heat of the bush in order to cultivate an organism that evolved in the rainforest: a genus of fungus known as *Termitomyces*. The termites, in other words, are farmers. These fungi are specialists at breaking down lignin, disrupting the tough carbon lattice and converting it into a form that the termites can eat. The termites don't eat the fungus itself. Instead, the fungus grows on and breaks down the termites' faecal pellets, which contain large amounts of undigested lignin. These pellets, called fungal cones, are visible as small white structures within the mound. After a few weeks of processing, the fungi have degraded the fungal cones into a form that the termites can digest, and they are re-eaten.

This complex symbiotic relationship might seem like a precarious environmental niche for fungi and termites alike, but it is carbon cycling on an industrial scale: Around 90 per cent of the lignin in this part of Africa is returned to the food chain by these tiny farmers.

Once the lignin is unlocked, the flow of carbon takes on a more familiar path. Aardvarks (*Orycteropus afer*) feed almost exclusively on termites, and are in turn prey for lions, leopards and hyenas.

**BELOW:** The food chain continues, with termites being eaten by aardvarks, which eat virtually nothing else. In turn, aardvarks are eaten by other animals higher up the food chain.

**BELOW RIGHT:** The rumen is the primary site for microbial fermentation of ingested food in ruminants such as cows and deer. In this scanning electron micrograph, anaerobic bacteria (red) are in the process of digesting cellulose plant material (grey).

*Aardvarks feed almost exclusively on termites, and are in turn prey for lions, leopards and hyenas.*

The sprawling food web of the savannah isn't entirely dependent on intermediate aardvarks, of course. While the termites and their fungal crop are the kings of lignin digestion, cellulose takes a more direct route into the food chain through Africa's iconic herds of herbivores: the giraffe, bison, zebra and gazelle. All these animals face the same problem as the termites – they lack enzymes efficient enough to break into the tough carbon lattice of lignin and cellulose – and their solution is similar. They have turned to the biochemical alchemy of another kingdom of life – in this case primarily the bacteria. All these animals are ruminants, and have complex, four-compartment stomachs (see diagram). The last of these compartments, the abomasum, corresponds to our stomach. The rumen acts as a large fermentation chamber filled primarily with bacteria, but also protozoa, which possess the enzymes necessary to break down cellulose. The complexity is quite daunting, but absolutely necessary if you live primarily on grass. Even with this precision anatomy, however, these animals must eat for at least 60 per cent of the day to access the energy contained in these near-impenetrable chains of carbon.

And, just like the termites, as soon as you crack open the carbon chains, you too become a perfect source of nourishment. With the hard work done, it's the great, glamorous predators who stalk the African night looking for a free lunch. ◉

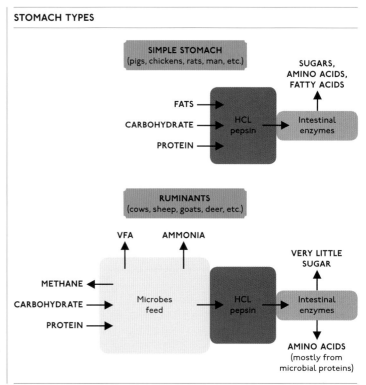

**STOMACH TYPES**

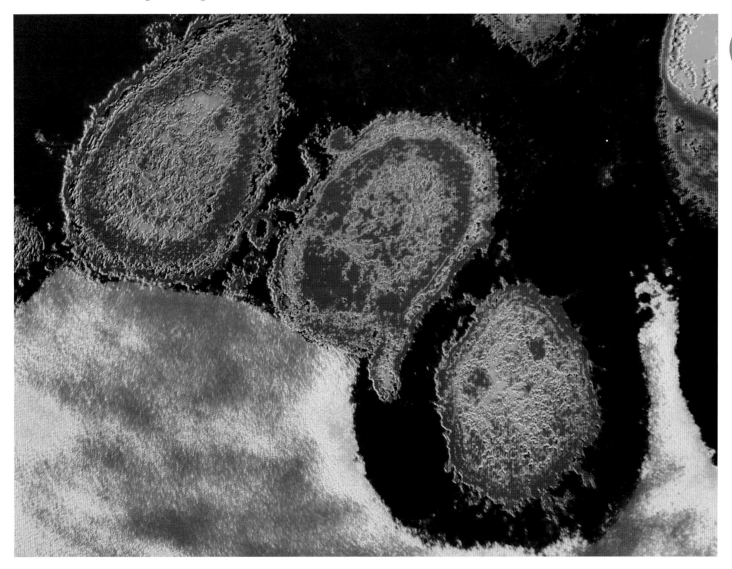

# WHY CARBON?

From hydrogen to plutonium, there are 94 naturally occurring elements on our planet, each with unique physical properties. Yet only a handful are found in every living thing, and carbon in particular seems to play a central, structural role in all biological molecules. What is it about carbon that gives it this pivotal position? The answer, as it must, lies in the chemistry of carbon, which in turn is dictated by the structure of the atom itself.

Carbon is the fourth most abundant element in the Universe after hydrogen, helium and oxygen. It has atomic number 6, which means that it has 6 protons in its nucleus. The most common form of carbon, known as carbon-12, has a nucleus that also contains 6 neutrons. Around the carbon nucleus there are 6 electrons, and it is the arrangement of these around the nucleus that determines the chemistry of carbon.

Electrons are subatomic particles, and their behaviour is described by quantum theory. A detailed description of the quantum theory of atoms is way beyond the scope of this book, but there are a few principles that will allow us to understand the behaviour of carbon atoms a little better. Most importantly, electrons cannot all crowd around the atomic nucleus in blissful ignorance of each other. They must obey something called the Pauli exclusion principle, which says that no two electrons (of the same spin, if we are to be precise) can be in the 'same place' (for the sake of precision again, we should really say in the same quantum state). The important consequence of this is that the electrons around the nucleus occupy different energy levels, or shells; you might imagine a series of available slots around the nucleus, each of which can hold a maximum of two electrons (with differing spins). The slot closest to the nucleus is called the 'K' shell, or sometimes the '1s' shell. This is the lowest energy slot, and two electrons can sit comfortably inside. Next is the 'L', shell, which has two sub-shells known as '2s' and '2p'. If you're interested in the meaning of the names, then have a look at any book on quantum theory. The key point is that there are four available slots in the 'L' shell, each of which could hold 2 electrons. In carbon, there are only 4 electrons in the 'L' shell, and they take up single occupancy positions in the four slots. This structure is sketched, highly schematically, in the figure below, but it is important to realise that atoms don't look like this in reality. The electrons are not in 'orbits'; rather they exist in intricately shaped clouds around the nucleus.

Carbon is able to form complex molecules because each of these 4 'L' shell electrons would like to pair up with an electron from a neighbouring atom, if at all possible. This is again loose language, but the principle is sound. If four hydrogen atoms are in the vicinity of the carbon atom, for example, then the single electrons around the four hydrogen nuclei can be shared with carbon's four available electrons to form a molecule of methane, $CH_4$. Carbon is also extremely happy to share electrons with other carbon atoms. Ethane, for example, consists of two carbons sharing one of their 'L' shell electrons with each other, with the remaining 6 electrons pairing up with hydrogen's to form $C_2H_6$. Hopefully you can see that carbon is the ideal building block, with an almost infinite appetite for sharing its outer electrons with any atom that will play ball.

There are other elements that also contain 4 solitary electrons in their outer shells. The next lightest is silicon, with 14 protons and 14 neutrons in its nucleus. It has a full 'K' shell, a full 'L' shell, and 4 electrons in the next-highest 'M' shell. Again, it is worth emphasising that the reason for this distribution of electrons around the nucleus can be found in quantum theory – it is not random, but dictated by the laws of physics. Silicon, therefore, has chemical behaviour similar to that of carbon. It, too, will form a molecule with four hydrogen atoms; $SiH_4$. This is known as Silane. It also forms $Si_2H_6$, which is known as disilane. And so on. In principle, therefore, silicon might be expected to be able to replace carbon as the scaffolding for the myriad molecules of life. The problem, however, is that silicon reacts differently to carbon. The reason is that silicon is a larger atom, and its full inner 'L' shell alters the chemistry significantly. The long-chain Silanes, for example, in contrast to long chain hydrocarbons, are not stable in water.

Despite their similarities, then, silicon and carbon have very different chemistries, and it is thought unlikely that silicon could replace carbon, perhaps on some alien world, as the molecule at the heart of life. ◉

---

**THE CARBON AND SILICON ATOMIC STRUCTURES**

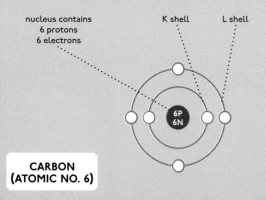

nucleus contains
6 protons
6 electrons

K shell    L shell

6P
6N

**CARBON
(ATOMIC NO. 6)**

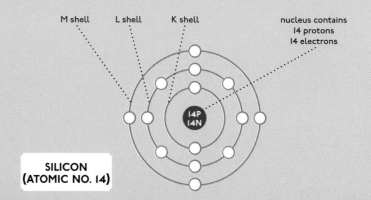

M shell    L shell    K shell

nucleus contains
14 protons
14 electrons

14P
14N

**SILICON
(ATOMIC NO. 14)**

## DIAMOND

## BUCHERENE

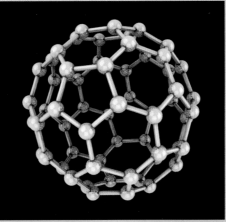

## GRAPHITE

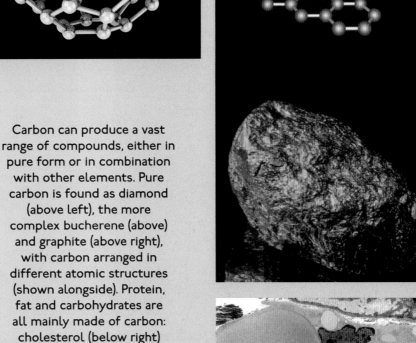

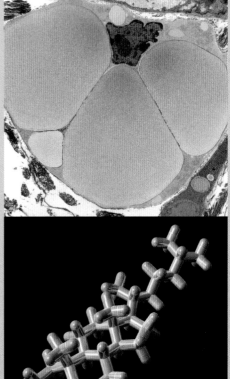

Carbon can produce a vast range of compounds, either in pure form or in combination with other elements. Pure carbon is found as diamond (above left), the more complex bucherene (above) and graphite (above right), with carbon arranged in different atomic structures (shown alongside). Protein, fat and carbohydrates are all mainly made of carbon: cholesterol (below right) is made of carbon, hydrogen and oxygen, protein (below) has added nitrogen molecules and glucose (below left) is, like cholesterol, also only composed of carbon, hydrogen and oxygen.

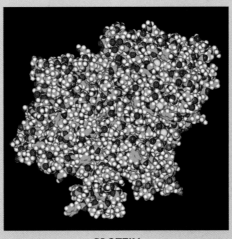

## GLUCOSE

## PROTEIN

## CHOLESTEROL

# THE BUILDING BLOCKS OF BIOLOGY

When Darwin discovered the fundamental processes that drive the constantly evolving tree of life, he knew very little about the underlying biochemistry – in particular, the hereditary mechanism by which information is passed from one generation to another. A handful of life's fundamental building blocks, the amino acids, were well known by the mid-nineteenth century. The first was isolated from asparagus in 1806, and is still known today as asparagine. Asparagine has the chemical formula $C_4H_8N_2O_3$, making it one of the simplest of the amino acids. All share a common basic structure, shown in the simplified diagram. The 'R' refers to the side chain, which differs in chemical structure for each amino acid. The simplest is alanine, whose side chain consists of a single carbon and three hydrogen atoms, known as a methyl group. There are 22 'standard' amino acids, of which 21 are used by eukaryotic cells. The other amino acid, pyrrolysine, is found only in archaea and a single species of bacterium. All amino acids contain four chemical elements: carbon, hydrogen, oxygen and nitrogen. Some also contain sulphur. These five elements might therefore be regarded as vital for life, or certainly life on Earth.

Amino acids are the building blocks of proteins, which form the familiar structures of all living things. Your skin, hair, muscles and tendons, the rhodopsin in your eyes and the haemoglobin in your blood, are all either made of proteins, or are simply proteins in their own right. Your enzymes, antibodies and hormones are also proteins. Some are even household names, such as collagen and insulin. Everything, in other words, from your structure to your biochemical functions, is either constructed from or operated by proteins.

Proteins can be extremely complex molecules, whose functions are determined not only by their chemical structure, but also by the way they fold. Discovering the three-dimensional structure of proteins is a very complex problem and is an important research area today, with widespread applications in medicine, as well as in the basic understanding of the processes of life.

The construction of proteins from their amino acid building blocks lies at the very heart of life, and the information to do this is carried by DNA. The details of how the information contained in DNA is used to build proteins are quite fiendishly complicated, but the principle is simple and elegant. DNA, or deoxyribonucleic acid to give the molecule its full name, is composed of two long chains of four simpler molecules known as bases, attached to a backbone of sugars and wound up into the famous double-helix structure discovered by Watson and Crick in 1953. The four bases are called adenine (A), cytosine (C), guanine (G) and thymine (T). Sequences of three bases code for amino acids; asparagine is coded for by AAT and AAC, and alanine is coded for by GCT, GCC, GCA and GCG. The reasons for the redundancy in the code are fascinating and not fully understood. In the above codes for asparagine and alanine it is the third letter that is

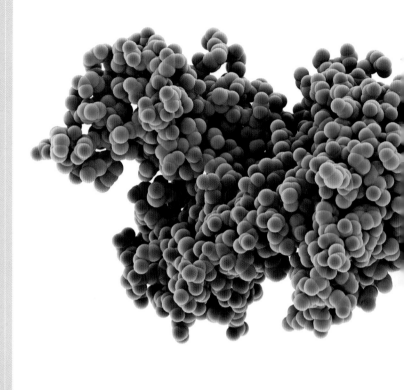

## THE BASIC STRUCTURE OF AMINO ACIDS
The R is known as the side chain. In the simplest amino acid – alanine – this consists simply of a methyl group, CH3.

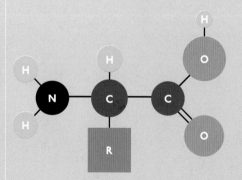

**TOP:** Molecular model showing a human T-cell receptor and a white blood cell antigen bound to a TAX peptide from a virus.

**RIGHT:** Computer model showing the structure of a molecule of protective antigen produced by anthrax ('Bacillus anthracis') bacteria. Anthrax employs a suite of three proteins, collectively termed anthrax toxin, that attack target organisms. Protective antigen is the first of these and works by providing a pore (hole, centre) in its victim's cell membrane, through which the other two proteins (not shown) can be delivered into the cell's interior.

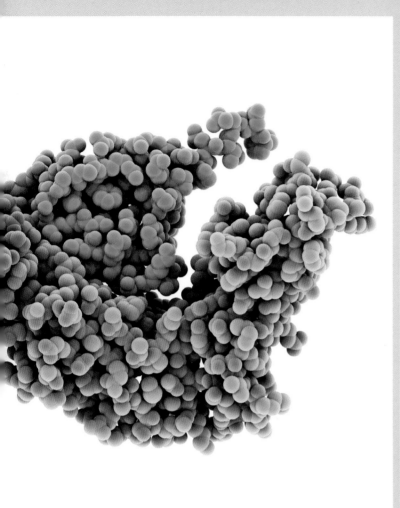

relatively redundant – a mutation in the third base often does not alter the amino acid that is produced; it is a 'silent' mutation that cannot be seen by natural selection.

Scientists do not know how or why the genetic code evolved, and there are several competing theories. Over 40 years ago, Francis Crick wondered whether it might not be a 'frozen accident' – a relic of our deep evolutionary past that just happens to be that way and which cannot change. There are now enough tantalising facts known about the code to convince many scientists that the code is not an 'accident', but they do not agree on why it is the way it is.

Some scientists think that life's original code may have used only two letters, coding for 16 amino acids; others suggest that the original code used four letters. Whatever the case, it appears that the first letter codes may be related to the way a particular amino acid is built by the cell, while the second letter may be correlated with how soluble the amino acid is. These kinds of findings are fascinating, suggesting that the genetic code is not simply an arbitrary information carrier. It looks as though the code may be intimately related to the physical chemistry of the amino acids themselves, in a way we do not understand. But this intriguing subject takes us beyond the scope of this book – for our purposes, the important thing is that these triplet sequences of bases code for amino acids, and the ordering of these triplets in DNA dictates how the amino acids are assembled into proteins. These sequences of triplet bases, coding for specific combinations of amino acids, are referred to as genes.

In summary, organisms are essentially collections of proteins, which form their structure and carry out all the complex biochemical functions of life. Proteins are built up from sequences of amino acids, and these sequences are coded for via the triplet code of DNA. There is a series of complex steps involved in going from the code to proteins, known as the central dogma of molecular biology: 'DNA makes RNA makes proteins'. RNA is a nucleic acid similar in structure to DNA, but single stranded and using the base uracil rather than thymine. But these details need not concern us here. The key point is that Darwin's hereditary mechanism is the DNA code, and that DNA code is converted into proteins in a series of chemical processes.

We will return to the genetic code, and its relevance to natural selection, later on. But it is worth looking again at those chemical elements that build up the amino acids: carbon, oxygen, hydrogen, nitrogen and sulphur, plus phosphorus (a component of DNA, RNA and ATP). These six elements are probably the minimal set necessary for life, at least in the form we find it on Earth. The scaffolding around which all biological molecules are constructed is carbon, which allows for the formation of long chains of hundreds or thousands of atoms because it is able to bond to four other atoms, including other carbons. The diagram of the structure of amino acids shows two carbon atoms, each forming four chemical bonds, sitting at their heart. The story of the origin of carbon in the Universe is worth exploring in more detail. We know that there was no carbon, or indeed oxygen, sulphur or phosphorus, present during the first few hundred million years in the life of the Universe. And for carbon in particular, its presence in large quantities may be a fortunate accident.

**RIGHT:** Like Brian, and all humans, and mammals such as the African lion, proteins have diverse roles in our bodies – they come in many shapes and sizes and are crucial to our existence. Hair and fur is made from a protein called keratin, the haemoglobin protein carries oxygen in our blood to every part of our body. Enzymes in our saliva and stomach are proteins that help digest our food, and muscle proteins called actin and myosin enable us to move – from breathing and walking to blinking our eyes.

ENDLESS FORMS MOST BEAUTIFUL

# DNA FROM THE BEGINNING

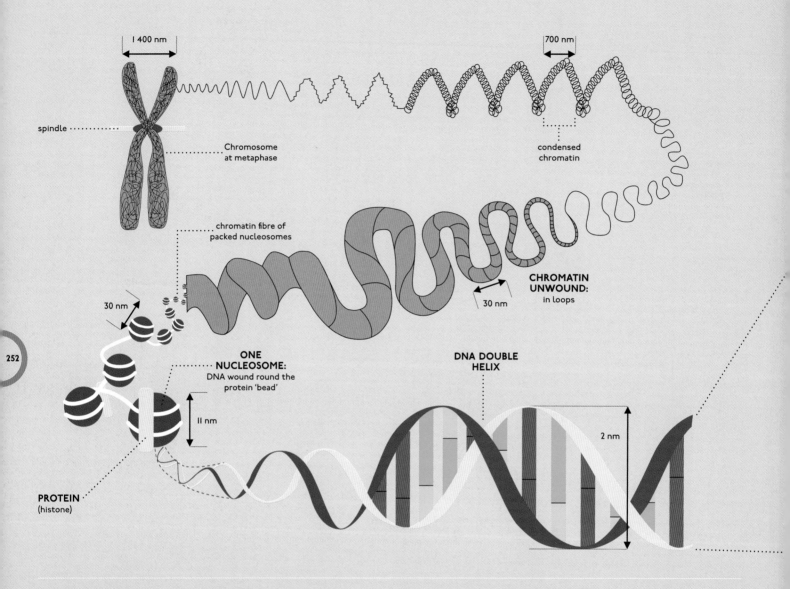

**1 400 nm**

**700 nm**

spindle

Chromosome
at metaphase

condensed
chromatin

chromatin fibre of
packed nucleosomes

**CHROMATIN
UNWOUND:**
in loops

30 nm

30 nm

**ONE
NUCLEOSOME:**
DNA wound round the
protein 'bead'

**DNA DOUBLE
HELIX**

11 nm

2 nm

**PROTEIN**
(histone)

Look in detail at the structure of DNA and you find a molecule of beautiful simplicity. Deoxyribonucleic acid is a molecule constructed from carbon, oxygen, nitrogen, hydrogen and phosphorus; a recipe that in the way it structures itself creates a code that can contain the instructions for every life form on Earth. It sounds as if this code should be dripping with complexity, but the framework it is created from is a lesson in biochemical minimalism. The backbone of the molecule is constructed from a sugar called 2-deoxyribose, which is a five-carbon sugar that is joined together in long strands by linking up with a phosphate group. This phosphate-deoxyribose backbone contains none of the vital information DNA must convey, but it does provide the framework on which the code can be built, and this is found in the four types of molecule that attach themselves to the backbone. It is a code written in a series of chemical groups called nucleotides that creates

the whole language of life with just the letters ATGC. These nucleotides, or bases as they are commonly known, are abbreviations of adenine, cytosine, guanine and thymine and are classified into two groups: the purine molecules A and G, and the pyrimidine molecules C and T. When attached to the deoxyribose backbone, these bases form long polymer chains and, crucially, each DNA molecule is made up of two of these chains. It is this that is the clever bit because the specific nucleotides in each group can only interact with one other specific nucleotide. It is called base pairing and it means that A can bond only to T, and G can bond only to C, and the result is that the strands of the DNA molecule are constructed in a perfectly complementary fashion. Crucially, the bonds between the bases on the two strands are not tightly locked covalent bonds that hold the rest of the structure together; rather, they are a weaker kind of bond called a hydrogen bond. The result of

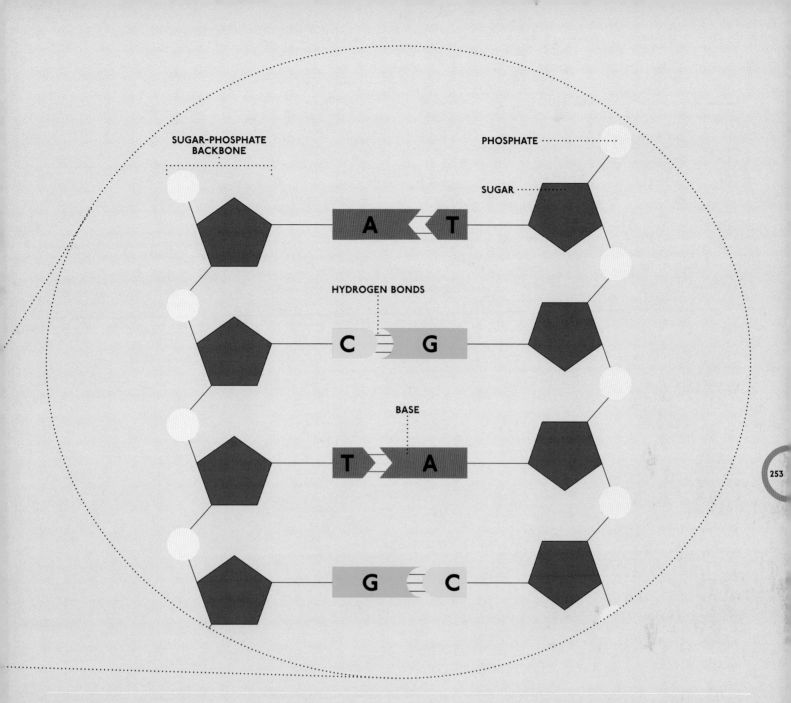

**SUGAR-PHOSPHATE BACKBONE**

**PHOSPHATE**

**SUGAR**

**A** **T**

**HYDROGEN BONDS**

**C** **G**

**BASE**

**T** **A**

**G** **C**

this configuration is where the real magic of DNA lies. As Watson and Crick put it in a piece of classic understatement in 1953, when they first described the double-helix structure of DNA: 'It has not escaped our notice that the specific pairing we have postulated immediately suggests a possible copying mechanism for the genetic material.' Just take a moment to think about a single fact – the fact that each of the trillion or so cells in your body was created from a single cell that formed at the moment of your conception. That original DNA in that initial cell of you is now in every cell of your body, and to do that has required a process of replication and division on a mind-boggling scale. Each time one cell turns into two it requires the DNA to be copied, and the way it does this is by exploiting the two strands that make up the molecule. Its design enables the entire structure of DNA to be unzipped perfectly down the centre – a division that opens the door for two new

identical DNA molecules to be made. With the molecule split apart, a cell can rebuild each strand only by using the complementary pairing A to T and G to C, and so a cell can rebuild itself only to exactly match the original strand of DNA. This process, known as mitosis, underpins the replication of every DNA molecule on the planet, allowing two almost identical molecules of DNA to be created from one. It is a physical process that not only directly links you to your parents and grandparents and beyond, but also much further back – back to a time of our distant primate ancestors, back to a time of the first mammal, the first vertebrate, the first multicellular organism, and back even earlier than that. The amazing thing about the DNA inside every cell of your body is that it connects in an unbroken line all the way back through almost 4 billion years of time to the very first life that used the remarkable, adaptable and enduring carbon-based molecule that is DNA. ◉

# MEET THE ANCESTORS

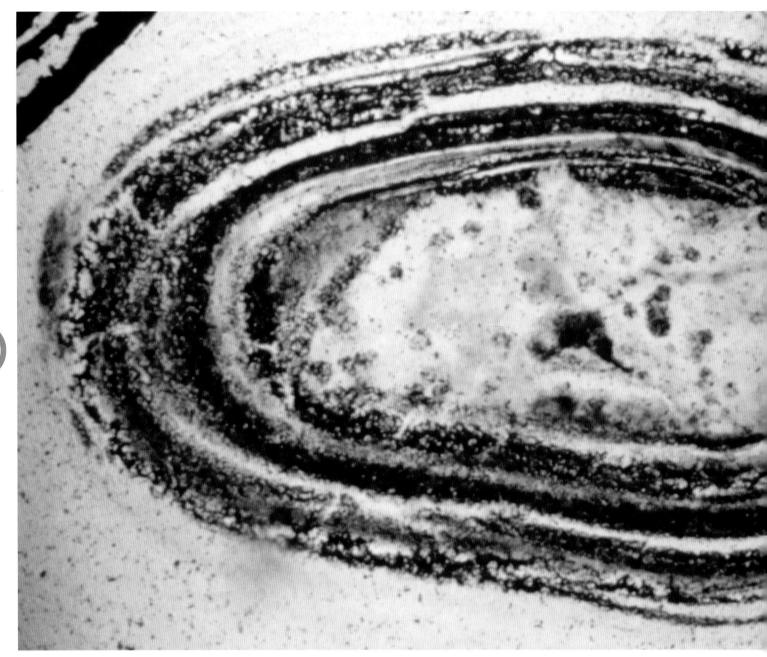

In the first half of this chapter, we have glimpsed the complexity and diversity of life on the African savannah and in the forests of Madagascar. We have also considered the biochemical underpinnings of life, in the form of amino acids, proteins and DNA, and followed carbon on its long journey from the stars and into Earth's food chain.

Darwin knew very little about this molecular world below the skin; he was primarily concerned with the origin of diversity – the origin of species – given these underlying mechanisms. The ability to focus on a specific part of an immensely complex problem is one of the hallmarks of a great scientist. Questions of the origin of life, the chemical and

biological machinery of animals and plants, and the history of our planet and all the living things upon it would have been utterly intractable in the mid-nineteenth century, and if Darwin had worried about solving the problem as a whole he would never have made progress. Instead, Darwin focused on the very specific question posed in the title of his famous work: what is the origin of species? In modern language, what natural processes caused, or allowed, a simple population of biochemical replicators to explode into the endless species we see on Earth today?

Let us begin with an original population of organisms – the foundation of all life on Earth today. They are collectively

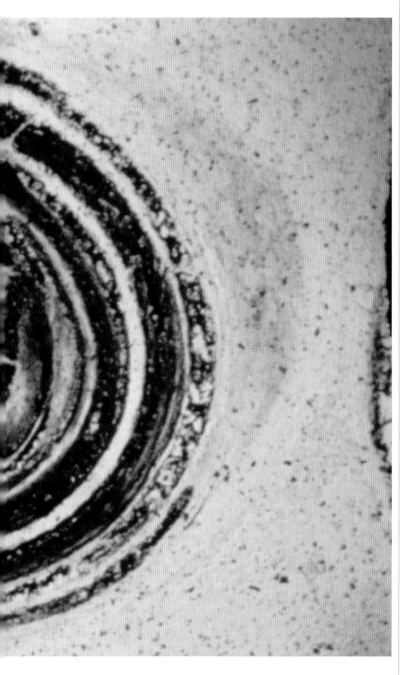

**BELOW:** Microfossil remains of the earliest life forms yet found, from the Gunflint cherts in Ontario, Canada, and dating from around 2 billion years ago. But how far back in time can we go in the search for LUCA – the Last Universal Common Ancestor?

known as LUCA – the last universal common ancestor. These words mean something quite specific and dizzying. If you trace your genetic lineage back through the centuries, from your parents, grandparents, great grandparents, and so on, then we know that you will arrive at a small group of hominid ancestors living in the vicinity of the Great Rift Valley around modern-day Ethiopia and Tanzania around 2 million years ago. Sweep back up your family tree still further, and you will pass all the familiar landmarks we've seen throughout this book. There will be shrew-like mammals, amphibians, fish, the first vertebrates, the first eukaryotes, and so on, and you are directly related to them all. There must be an unbroken line, stretching

back 3.8 billion years or more, from you, to LUCA. Every single one of your relatives lived long enough to produce offspring; not a single one died before passing on their genetic code.

As we saw in Chapter 2, it is possible that LUCA was not even a cell, but a collection of biochemical reactions involving proteins and self-replicating molecules. Some biologists believe that before LUCA there was another kind of replicative chemistry that involved RNA alone, as RNA can replicate without the complex enzymes that allow DNA to replicate inside living cells today. But when we speak of LUCA, the details of the origin of life don't really concern us. What matters is that there was a set of molecules that encoded information that could be passed down from generation to generation. This is what Darwin meant by a 'primordial form, into which life was first breathed'. The task he set himself in *On the Origin of Species* was to explain how this population of replicators gave rise to the millions of species alive today.

### THE GREAT GENETIC DATABASE OF LIFE

We begin, then, with a population of living things equipped with DNA and the minimal set of enzymes and biochemistry required to allow the organisms to replicate. To reiterate an important point, all life today shares the same genetic code, and this is one of the key pieces of evidence that suggests a common origin in the population, known as LUCA. The information contained in the DNA of this population can be seen as a database, spread over many individuals no doubt, but still localised in one particular form of life. All the information necessary to build new members of this population must have been held in this database.

Now, 3.8 billion years later, the database has expanded almost beyond comprehension, and it is constantly changing as new organisms are born, and others die. Encoded within it is information about the history of our planet, with its dramatically shifting geology and environment, because these changes have affected the form and function of living things. It also documents the interactions of organisms with one another. As we have seen, the structure of Darwin's orchid is impossible to understand without reference to the *Xanthopan morgani praedicta* moth, so the database contains not only the information to construct each, but also information about their joint historical development. Information has been lost in extinctions; the great database no longer contains the complete information necessary to build a *Tyrannosaurus rex*, but information about dinosaurs will still be there in the imprints their interactions left on extant species, and in their feathered descendants, the birds. In its entirety, then, this grand genetic database contains all the information necessary to build every organism alive today, and what it took for the ancestors of those organisms to survive in every niche and through every epoch.

The challenge for any theory that purports to explain the diversity of life on Earth is to explain how this DNA database came to be so diverse, and crucially how it came to be separated and spread among millions and millions of different organisms, each possessing enough information to construct itself, but no single one containing it all. Once, long ago, the entire database was held in the population known as LUCA. Now, it is fragmented across tens of millions of species. Why? ◉

255

WONDERS OF LIFE

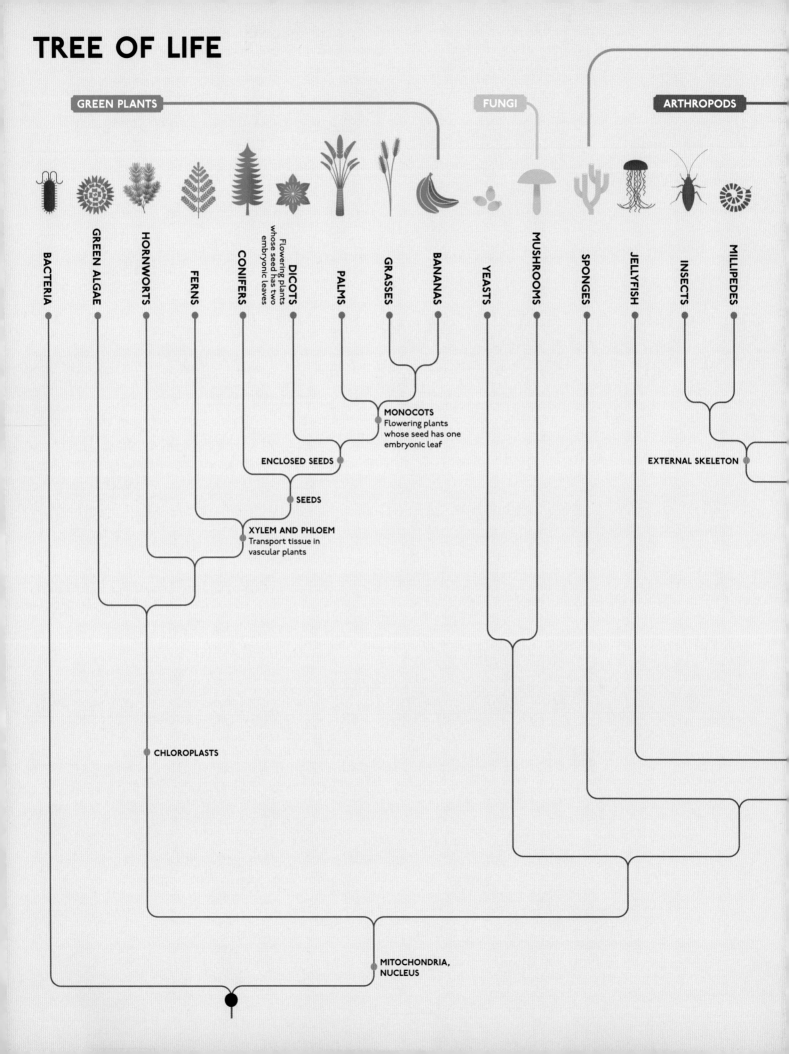

# TREE OF LIFE

GREEN PLANTS

FUNGI

ARTHROPODS

BACTERIA

GREEN ALGAE

HORNWORTS

FERNS

CONIFERS

DICOTS
Flowering plants whose seed has two embryonic leaves

PALMS

GRASSES

BANANAS

YEASTS

MUSHROOMS

SPONGES

JELLYFISH

INSECTS

MILLIPEDES

MONOCOTS
Flowering plants whose seed has one embryonic leaf

ENCLOSED SEEDS

SEEDS

XYLEM AND PHLOEM
Transport tissue in vascular plants

EXTERNAL SKELETON

CHLOROPLASTS

MITOCHONDRIA, NUCLEUS

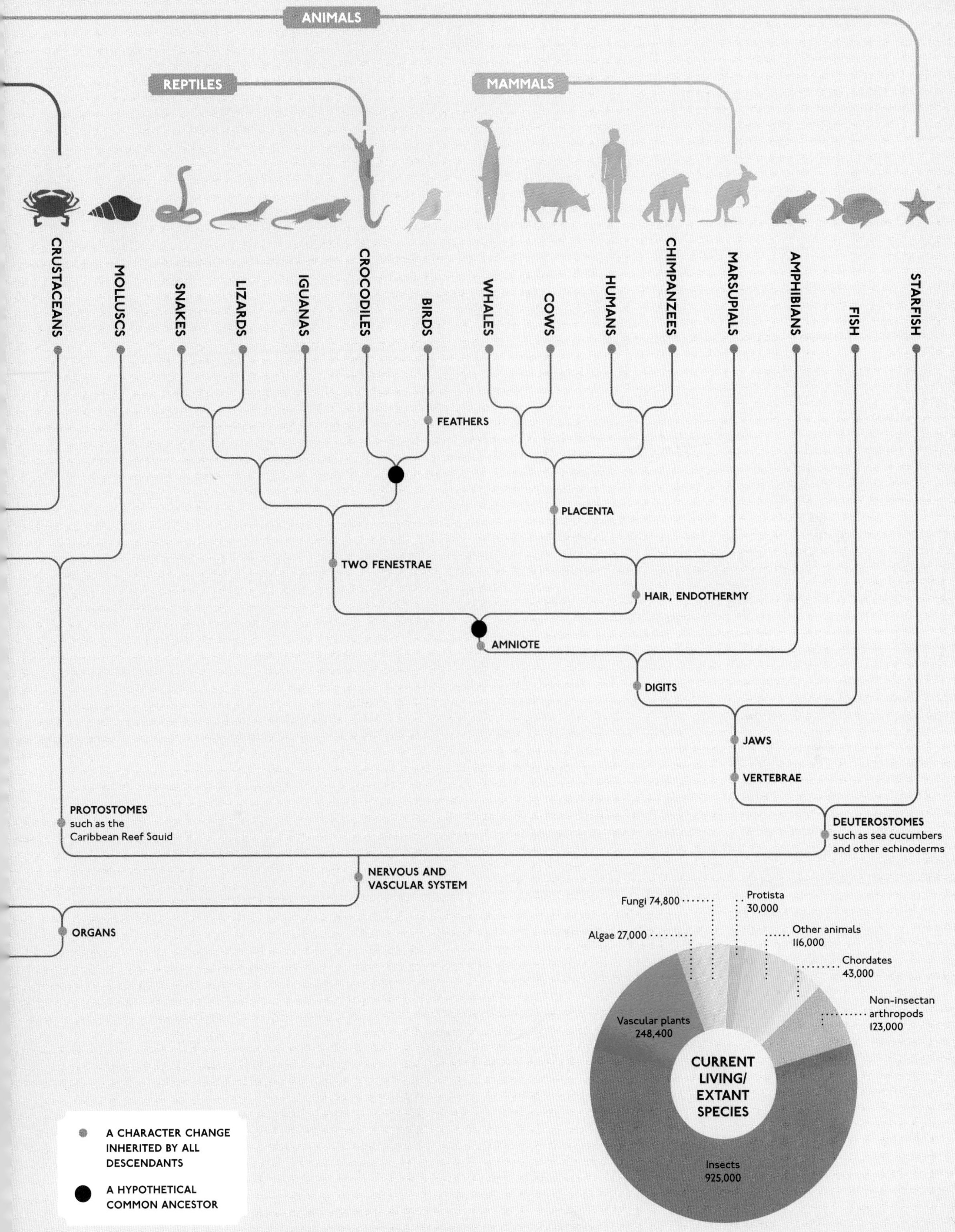

ANIMALS

REPTILES

MAMMALS

CRUSTACEANS
MOLLUSCS
SNAKES
LIZARDS
IGUANAS
CROCODILES
BIRDS
WHALES
COWS
HUMANS
CHIMPANZEES
MARSUPIALS
AMPHIBIANS
FISH
STARFISH

FEATHERS

PLACENTA

TWO FENESTRAE

HAIR, ENDOTHERMY

AMNIOTE

DIGITS

JAWS

VERTEBRAE

PROTOSTOMES
such as the
Caribbean Reef Squid

DEUTEROSTOMES
such as sea cucumbers
and other echinoderms

NERVOUS AND
VASCULAR SYSTEM

ORGANS

A CHARACTER CHANGE
INHERITED BY ALL
DESCENDANTS

A HYPOTHETICAL
COMMON ANCESTOR

Fungi 74,800

Protista
30,000

Algae 27,000

Other animals
116,000

Chordates
43,000

Vascular plants
248,400

Non-insectan
arthropods
123,000

CURRENT
LIVING/
EXTANT
SPECIES

Insects
925,000

# MUTATIONS: THE SPRING FROM WHICH DIVERSITY FLOWS

I magine the ancient sequence of bases in the DNA of LUCA. We have no idea how long these sequences were. Today, the smallest known genome of any free-living organism is that of a bacterium known as *Pelagibacter ubique*, which contains just under 1.5 million base pairs. Even this compact genome is enough to code for all 20 of the amino acids. The smallest genome capable of coding for a self-sufficient organism is unknown; theoretical estimates suggest somewhere in the region of a quarter to half a million base pairs, which is of similar length to known parasitic organisms.

If LUCA were a perfect replicator, copying its sequence dutifully and without errors or changes from generation to generation, then nothing would change and the world today would still be populated by 'LUCAs', assuming they survived the inevitable changes in their environment. There are, however, regular and inevitable changes in the sequence, which occur during copying and through external causes. These changes are the wellspring from which diversity flows. Let us focus on one particular source of change – random changes in a single letter in the sequence of bases caused by external factors. These are known as point mutations, and for the purposes of our film, we chose to focus on a particular and important source of these mutations that comes from a rather surprising and exotic source.

## CHANGE FROM THE HEAVENS

Cosmic rays are high-energy subatomic particles that bombard the Earth from space. Around 90 per cent are protons, the nuclei of hydrogen, and most of the remainder are helium nuclei. Some of the lower-energy cosmic rays originate on the Sun, but the highest-energy particles have an unknown origin far outside our Solar System, and indeed beyond our galaxy. It is thought that supermassive black holes at the centre of active galaxies are the source of the most energetic cosmics. The most infamous of these was the 'Oh My God!' particle observed over Utah on 15 October 1991 – a single proton that carried the energy of a tennis ball served by a Wimbledon champion. If this sounds like a science-fiction source of genetic mutations, then note that research by IBM has suggested that every computer and mobile phone on the planet suffers on average one error per 256 MB of memory per month as a result of cosmic ray strikes. This means that the information in your DNA is suffering a similar fate. In fact, over 10 per cent of the natural background radiation on Earth at sea level comes from cosmic rays, and this rises dramatically with altitude.

Allowing myself an aside, it is worth noting that the American physicist Carl Anderson discovered antimatter in a particle collision initiated by a cosmic ray in 1932. Until

*Genes are shuffled by sex, there are copying errors introduced as the code is copied and passed from generation to generation...*

the advent of particle accelerators, cosmic rays were the only source of high-energy particle collisions, and we will never have the technology to accelerate particles to the energies seen in the most violent cosmic-ray collisions. The Large Hadron Collider is a toy compared to a supermassive black hole and the magnetic fields of a galaxy of a trillion stars. When we were in the process of filming our cosmic-ray sequence, I mused upon whether some kind of key mutation that led to the emergence of the human species might have been triggered by a proton, accelerated to near light-speed by a distant black hole, smashing into the DNA of an unsuspecting cell in the body of a primitive animal in a primordial ocean. Unlikely, but you can always dream.

Cosmic rays are a significant source of mutations. Natural background radiation from radioactive isotopes in rocks, the action of chemicals and ultraviolet light from the Sun are others. Genes are shuffled by sex, there are copying errors introduced as the code is copied and passed from generation to generation, and whole sequences of code can be transferred from one species to another in a process known as lateral gene transfer. This is the origin of the spread of antibiotic resistance in bacteria.

Together, these random changes pose a significant challenge to life's repair mechanisms. In total, there are around a million instances of molecular damage in your body every day from external sources, and a significant fraction of these are damage to your DNA. Cell death or cancer can result. The overwhelming majority of these damaging events are corrected by the body's in-built repair mechanisms, but the success rate is not 100 per cent, and if uncorrected damage occurs to the DNA contained within an egg or sperm cell, then the resulting mutation may be passed on to the next generation. ◉

# THE POWER OF MUTATIONS

For a relatively insignificant small fly with a taste for rotting fruit, the *Drosophila* has had a remarkable impact on scientific knowledge. One species in particular, *D. melanogaster*, has played a disproportionate role in twentieth-century human history. Much of the reason for this stems from the work of Hermann Muller, a New York-born geneticist who won the Nobel Prize for a series of experiments he carried out at the University of Texas in the 1920s. Muller wanted to investigate the effect of X-ray radiation on the mutation of genes, and he chose to explore this question using *D. melanogaster*. At the time, no direct causal link between radiation and genetic mutations had been demonstrated, but as Muller began to irradiate his flies with varying doses of X-radiation, it soon became apparent that a clear relationship existed. Muller irradiated generation after generation of *Drosophila*, and showed that there exists a strong correlation between radiation dose and the frequency of mutation. This was headline news to a generation who had grown up considering radiation as bordering on the almost wholesome. Many of the early pioneers of radioactivity died young of the unknown effects of exposure. Marie Curie, who died in 1934 at the age of 66 from radiation-related illness, wrote in notebooks that are to this day kept in lead-lined boxes and cannot be handled without modern radiation safety precautions being observed.

**RIGHT:** The eyes of a fruit fly (top) are naturally red, but the production of pigments in the eyes can be switched off, resulting in a fly with white eyes (bottom).

**BELOW RIGHT:** Hermann Muller demonstrated the strong link between radiation dose and the frequency of mutation, and showed that detrimental mutations are far more common than beneficial ones. This image shows *Drosophila melanogaster* carrying the 'stubble' and 'drop' mutations.

**BELOW:** The fruit fly (*Drosophila melanogaster*) was the principal focus of study for geneticist Hermann Muller. Here, the *Antennapedia* mutant has legs instead of antennae, caused by genetic alteration.

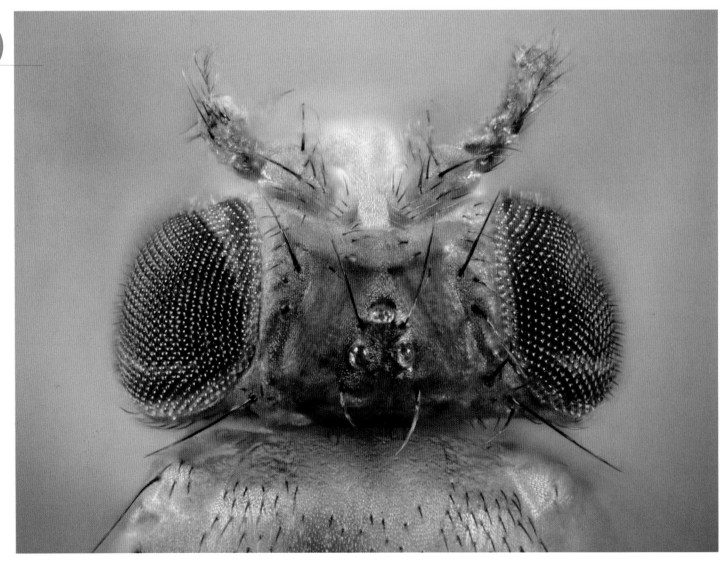

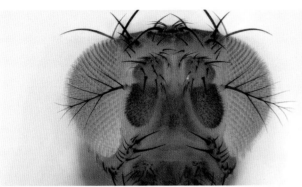

Muller's work, a decade before Curie's death, cemented the modern fear of radiation in the public mind, and in doing so played a part in transforming the political, technological and environmental landscape of the twentieth century. Today, the word 'radiation' is almost synonymous with mutation.

Muller observed that the majority of mutations generated by X-rays were detrimental. While mutations may throw up the occasional 'good gene', Muller and his team were the first to show that detrimental mutations are far more common. A more recent study suggests that if a genetic mutation does change a protein, it will be harmful around 70 per cent of the time, with the remaining 30 per cent of changes being either neutral or only weakly beneficial.

Some of the less deadly mutations in fruit flies are still easy to spot. *Drosophila* eyes are naturally red, but with just the right mutation, the production of pigments in the eye can be switched off, resulting in a white eye. White-eyed flies are at a substantial disadvantage – they do not mate as successfully as red-eyed flies because they cannot see well and other flies show abnormal responses to them.

Despite the odds being stacked heavily in one direction, Muller was quick to realise the implications of this observation. Mutation could be a powerful force for change, and in a Universe full of radiation, there is no shortage of opportunity. ◉

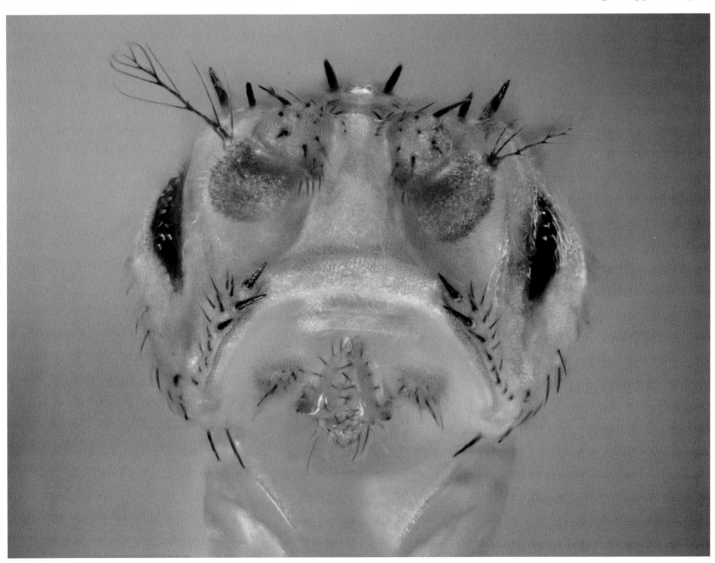

Changes to the genetic sequence, then, are common. They are happening all the time, in every living thing, and provide a rich source of change. They are the source of what Darwin called variability, the wellspring of diversity. But they cannot alone explain the diversity of life on Earth. To see why this is so, we need to do a little bit of simple statistics.

Each position in the genetic code can be occupied by one of the four bases – A, T, C or G. There are four ways, therefore, of changing a single base at random. One will leave it unchanged, and three will result in a different sequence, which may code for a different amino acid. If two bases are changed at random, there will be 4×4 = 16 possibilities. Three changes result in 4×4×4 = 64 possibilities, and so on. This exponential growth in the number of combinations means that for a code containing only 150 bases, there are more possible combinations than there are atoms in the observable Universe. In other words, the chance of coming up with even the simplest genetic code in the simplest organism on Earth at random is vanishingly small. And considering that the human genome is over three billion bases long, you see the problem.

This is worth emphasising. ANY theory that explains the evolution and diversity of life must have a significant non-random component. There are simply too many combinations in the genetic code for anything interesting to appear spontaneously as a result of random mutations and assembly of bases. As Darwin's theory of evolution by natural selection does explain the diversity of life, then it should come as no surprise to hear that natural selection is, indeed, non-random.

Given variation, then, we are still faced with the seemingly intractable problem of statistics. Living things such as human beings are incredibly complicated, and it is impossible to get from LUCA to us in only 3.8 billion years simply by random changes in the genetic code. It would, in fact, be impossible in a universe a trillion times older than ours. There must be natural ways of narrowing down this vast landscape of genetic possibility, and indeed there are many and they are powerful.

One particularly effective selection effect is the requirement that every code must produce a living animal. This is highly restrictive, and rules out the overwhelming majority of possible codes. Death before

**OPPOSITE PAGE, LEFT:** Mutations that are detrimental – as with this two-headed albino California kingsnake ('Lampropeltis getula californiae') – are unlikely to live long, and are therefore unlikely to pass their genes on to the next generation.

**OPPOSITE PAGE, RIGHT:** Albinism – shown in this image of an American alligator ('Alligator mississippiensis') – is characterised by the lack of pigment in the skin, hair and eyes due to the absence or defect of tyrosinase, a copper-containing enzyme involved in the production of melanin.

**THIS PAGE:** Albino tapir, hedgehog and grey squirrel. Albinism can lead to increased susceptibility to infection, and consequently it is rare.

birth, or even the prevention of conception, is the most brutal example of natural selection. As Richard Dawkins memorably put it, 'however many ways there may be of being alive, it is certain that there are vastly more ways of being dead!'

From LUCA onwards, only sequences of bases that resulted in living things survived. This is a statement of the blindingly obvious, but it is the key to understanding natural selection. ALL random changes that result in death are removed from the landscape of possibilities. This is rule number one if you like – the most coarse-grained element of the sieve of natural selection.

Another brutal layer of the sieve is provided by mutations that lead to dramatic changes in an individual – so-called macromutations. Changes to the genes that control body plan might fall into this category. Individuals born with legs instead of antennae (such as the fruit fly on the previous page) for example, might live for a short time, but will almost certainly not pass this mutation on to the next generation.

The preceding two examples are both negative, in the sense that they guillotine a huge array of changes, dramatically restricting the landscape of possible genetic codes. These theoretically possible sequences of bases, which form the overwhelming majority of the set, will never be allowed to appear in the database of life. And so our statistics problem begins to seem less intractable, because of the simple fact that only those sequences of bases that produce viable organisms will be present in the great genetic database.

Changes can also be 'neutral' in the sense that they have no effect on the fitness of an organism – this is particularly the case for mutations in those substantial parts of our genetic sequence that do not appear to code for anything, or, more interestingly, for mutations that produce a small quantitative variation in a particular character. Very rarely, changes improve the chances of an organism passing its genes on to the next generation. Precisely how natural selection acts on this constantly fizzing, shifting landscape of possibilities to produce the diversity of living forms we see today is the subject of the remainder of this chapter, and the ideal place to let the story unfold is back where we started, on the magical island of Madagascar. ◉

# THE POWER OF
# ISLANDS

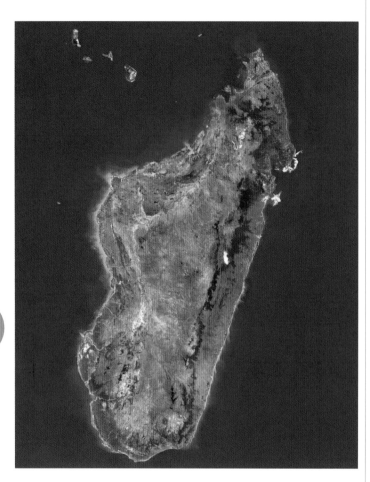

the settlers from the raft. Genetic studies show that either there was only one migration, or that if there was more than one, then they occurred closely together, and were from the same African population. The story of Madagascar's lemurs provides us with an excellent analogy for the evolution of life from LUCA to the present day. From the point of view of modern lemurs, the individuals on the raft are LUCA – a small population carrying with them a genetic database with the instructions necessary to build a lemuriform. How did that database expand and fragment to produce the diverse and highly specialised range of lemurs living in Madagascar today?

Before we meet the lemurs themselves, there is one central idea that can already be explored. Modern lemurs exist on Madagascar and nowhere else because Madagascar is an island. When our seafaring travellers landed on the shore so many million years ago, there were no other primates on Madagascar. They brought with them in their genetic sequence a limited subset of the great library of life – a selection of the books, if you like – containing the instructions to make a lemuriform. This is the raw material that mutation and natural selection worked upon, isolated and separated from mainland Africa, to produce the lemurs of today. This is radically different to the situation on the mainland, where there was an expanded library even among the lemuriforms themselves, because there were many more individuals. There were also different selection pressures, most notably competition from other primates, which ultimately led to the disappearance of the lemur ancestors from the mainland – although their close relatives, the lorises, survived.

The island of Madagascar itself, therefore, is the most important factor in the story of the lemurs. The physical separation and isolation of a subset of the genetic database of life allowed forms to emerge that are different from those on the mainland. The reasons for the differences should be obvious. The random mutations that occurred in this isolated gene pool on Madagascar never got transferred to the mainland populations. The sieve that selected for and against these mutations was also different; importantly, there was no competition from other primates, and the geological and natural history of Madagascar itself is, of course, different to that of mainland Africa. The surviving changes in the lemurs' genetic database were determined by natural selection, testing them against Madagascar, with its unique climate, flora and fauna. The Madagascan sieve is different to that of the mainland, and so the genetic database it produced will be different and therefore the animals that are an expression of that database will be different.

In summary, it is the isolation of specific parts of the great genetic database of life, and the subsequent cycle of mutation, mixing of genes, and the sieve of natural selection, that produces speciation – the divergence of forms from their distant ancestors that ultimately results in the development of new species.

Let us draw all these threads together in a single animal that is without doubt the rarest and strangest animal I had the good fortune to meet during the filming of *Wonders of Life*: the aye-aye. ◉

S omewhere between 50 and 80 million years ago, a handful of seafarers were nearing the end of a 560 km voyage across the Mozambique Channel. They were floating on a natural raft of vegetation that had been their home for weeks. They were accidental travellers; creatures from mainland Africa that had been trapped and taken by the ocean's currents. The land they found was rich in trees, plants and animals, but there was none of their kind. Fate brought a group of animals to the shores of an island and, over the course of time, those ancestral primates have evolved and diversified to become Madagascar's most iconic animals: lemurs.

Madagascar's lemurs provide a powerful example of Darwin's insight into the origin of species. They illustrate all the essential ideas behind Darwin's theory in an intellectually satisfying and visually beautiful way, and cement the island's place as the perfect location for the final act of our film.

Today, there are 99 living species and subspecies of lemur on Madagascar, none of which is found anywhere else on the planet. They form a quite dazzling array, from the beautiful to the bizarre, and they are all directly related to

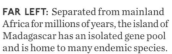

**FAR LEFT:** Separated from mainland Africa for millions of years, the island of Madagascar has an isolated gene pool and is home to many endemic species.

**LEFT:** All lemurs are endemic to Madagascar. This diademed sifaka (*Propithecus diadema*) is one of the largest and most distinctive.

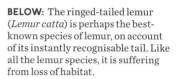

**BELOW:** The ringed-tailed lemur (*Lemur catta*) is perhaps the best-known species of lemur, on account of its instantly recognisable tail. Like all the lemur species, it is suffering from loss of habitat.

**MIDDLE:** The large-scale destruction of Madagascar's forests means that species such as the black lemur (*Eulemur macaco*) are becoming increasingly rare.

**ABOVE:** The random mutations that occurred over millions of years have resulted in a wide variety of lemur species. Compare this white-furred Von der Decken's sifaka (*Propithecus deckenii*) with the black lemur (above).

**RIGHT:** The crowned lemur (*Eulemur coronatus*) is endemic to northern Madagascar's dry deciduous forests.

265

# ISLAND MADAGASCAR

One hundred and seventy million years ago, Madagascar was landlocked in the middle of the supercontinent Gondwana, sandwiched between land that would eventually become South America and Africa and land that would eventually become India, Australia, and Antarctica. Through movements of the Earth's crust, Madagascar, along with India, first split from Africa and South America and then from Australia and Antarctica, and started heading north. India eventually collided with Asia, but Madagascar broke away from India and was left isolated in the Indian Ocean. Madagascar has been on its own for the past 88 million years.

## GEOLOGICAL HISTORY

### 170 MILLION YEARS AGO

### 162 MILLION YEARS AGO

### 135 MILLION YEARS AGO

### 88 MILLION YEARS AGO

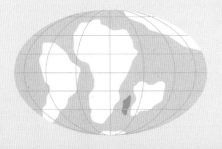

## BIRDS
As a large island with many different ecological niches, Madagascar is home to an amazing variety of birds. Of the 294 species recorded, 105 are found nowhere else on Earth.

## PALMS
Madagascar is home to a rich palm flora, of which only a few species are widespread outside the island.

## DARWIN'S MOTH
Darwin used the idea that adaptive traits can evolve by reciprocal selection ('co-evolution') to predict the existence of a moth that could pollinate an orchid that has a long nectar spur. He theorised there must be a moth whose proboscis was long enough to reach the nectar. The moth – 'Xanthopan morgani' – was eventually discovered after Darwin's death. The plant is known as 'Darwin's Orchid'.

## ORCHIDS
Madagascar is home to almost 1,000 orchid species in 57 genera, many of which are as endangered as the lemurs.

## LEMURS
The most famous of Madagascar's land mammals are the lemurs, who split off from the common ancestor of the bush babies after Madagascar became an island. They came by sea 60–50 million years ago, on large rafts of vegetation.

## CHAMELEONS
Madagascar is home to about half of the world's 150 or so species of chameleon.

## ENVIRONMENTAL DAMAGE
Madagascar is biologically one of the richest areas on Earth, and its plants and animals are among the most endangered. More than 90% of the original forest of Madagascar has been destroyed by logging, farming and mining, among other causes, resulting in habitat loss and endangering one of the rarest and most treasured ecosystems in the world.

## FOSSA

One of the seven endemic species of carnivore found on Madagascar, closely related to mongooses but with cat-like features, an example of convergent evolution.

## ISLAND EVOLUTION

### 1

An ancestral species colonises one island in a cluster of islands.

## TENRECS

Some tenrecs look remarkably like hedgehogs – another example of convergent evolution, as they are more closely related to aardvarks and elephants. Most of the 24 species of tenrec are endemic to Madagascar.

## BEELZEBUFO

Also known as the 'frog from hell', the 'Beelzebufo ampinga' lived around 65–70 million years ago. Fossil fragments show that it could have measured 20 cm across its squat head, and its body was probably more than 40 cm long.

### 2

The species then colonises the other islands in the cluster – or archipelago.

## FLIGHTLESS BIRDS

Massive herbivorous elephant birds – flightless ratites even larger than New Zealand moas – were once a common sight on the island, up until the 17th century.

### 3

Populations on different islands evolve to become different species.

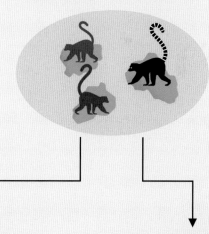

## FRESHWATER FISH

Madagascar's fish species are some of the most threatened on the planet. They include the Zonobe, the Menarambo and the Gogo catfish, among many others.

## PRESENT DAY

## GIANT TORTOISE

These giant reptiles are now extinct on Madagascar; the giant tortoises still found on the Seychelles and on Mauritius and Réunion were originally from Madagascar.

## ALLOPATRY

Species evolve while adapting to different environmental conditions on different islands, and in turn colonise other islands in the archipelago.

## SYMPATRY

Species colonise all the islands in the archipelago and then adapt to minimise competition with other species, which is called character displacement.

# A CREATURE FROM ANOTHER WORLD

The aye-aye is virtually demonic in appearance. This is not a trivial observation. One of the reasons it is critically endangered is that, historically, the aye-aye (*Daubentonia madagascariensis*) was seen as a bad omen and was killed on sight. Together with the destruction of its forest habitats, this has reduced the wild population to below 10,000 individuals.

Aye-ayes are the largest nocturnal primate, which makes these rare animals even more difficult to see in the wild. Indeed, for a time in the 1990s they were thought to be extinct. We were extremely fortunate in being allowed to film the capture, tagging and release of a wild aye-aye and her child by a research team led by Ed Louis from the Henry Doorly Zoo in Omaha. Tagging aye-ayes with GPS collars allows researchers to monitor their movements in order to better understand their population distribution and density, and develop strategies for their protection.

To say that capturing an aye-aye safely is not easy is an understatement of terrific proportions. The animal is an expert climber, and moves effortlessly through the high canopy of the forest. The forests themselves are the densest I have ever experienced, at times forming an impenetrable chaos of trees and bushes, many of which are threateningly spiked. And the darkness is impenetrable too – Madagascar's cloudy nights remove any chance of moonlight, and there are no cities to turn the sky a reflected orange. The only consolation is that Madagascar is devoid of venomous snakes and spiders, otherwise careering blindly through this snarled, scratching dark would have tested the nerves beyond reason.

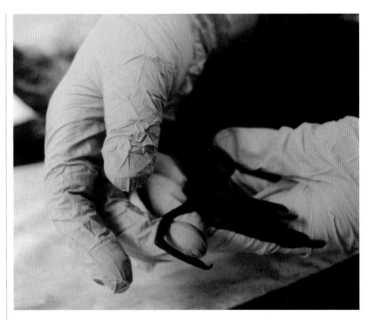

**TOP:** Madagascar is home to the critically endangered aye-aye (*Daubentonia madagascariensis*), a species of lemur that occupies a very particular kind of environmental niche – one that is occupied by woodpeckers in many other parts of the world

**ABOVE:** The aye-aye has an elongated and very flexible middle finger, which it uses first to tap the trunk of a tree in search of insects, and then to spear any grubs it finds and pull them out of the wood.

BELOW: Unlike all other primates, the aye-aye has continuously growing, forward-slanting teeth, rather like those of rodents, which it uses to chew through wood in search of grubs.

The aye-aye uses its finger to tap down the trunk of a tree, listening for a change in sound that might indicate the presence of a bug hiding inside. Our guide told me that they always tap three times. When a promising echo is located by those large, gremlinesque ears, they begin to gnaw through the wood. Aye-ayes are unique among primates in having continuously growing teeth. This trait, which is rodent-like, is vital, as their teeth are subjected to much greater wear and tear than those of their primarily vegetarian and insect-eating cousins. Once through the wood, the aye-aye deploys its finger again, spearing the grub inside the hole before levering it out to eat it.

This aye-aye's particular lifestyle explains its unique and startling form. It is nocturnal, so it has large eyes. It is a tree-dweller, which explains its dexterous hands and feet and large, counter-balancing tail. Its odd rodent-like teeth allow

*The aye-aye is the primate version of a woodpecker, deploying an utterly strange suite of adaptations in order to exploit the benefits of a wood-boring lifestyle.*

We stopped filming the hunt after around two hours, partly because we were exhausted and ripped to shreds, but more importantly because we felt we were in the way and disturbing both animals and trackers. It is not desirable to extend the hunt for too long, owing to the stress placed on the animals. Shortly after we left the forest, Ed's team managed to get a clear sight of the pair, tranquillise them both with darts, and catch the groggy aye-ayes as they fell safely down into carefully positioned nets.

When we arrived back at the camp in the damp morning, both animals had been tagged, given a thorough medical, and were sedated, ready for their release that evening. And marvellously, in one of my most exhilarating and precious experiences in the filming of all three series of 'Wonders', I was able to hold an aye-aye and use this strange animal to explain to a camera Darwin's theory of evolution by natural selection.

The aye-aye is a creature that lives in a very particular niche – one occupied by woodpeckers in many other parts of the world. There are no woodpeckers in Madagascar, which would have meant that grubs, buried deep inside the trunks of trees, could have led a rather charmed life immune from predation. The aye-aye is the primate version of a woodpecker, deploying an utterly strange suite of adaptations in order to exploit the benefits of a wood-boring lifestyle. The most startling is its unique middle finger, which is a grotesquely elongated, slender and bony structure with 360-degree movement on a ball-and-socket-like joint system. It feels broken as you gently rotate it around. The aye-aye's finger was longer than mine when fully extended, and this on an animal that is no bigger than a small dog.

it to gnaw through wood, and its finger – that strange central digit – allows it to access a readily available food source, safe from the competition of other large animals and birds. Virtually free from predation – only the fossa, a large, cat-like animal closely related to a mongoose, will pose any threat – the aye-aye has been free to specialise.

We can now draw all the threads together to describe how the aye-aye came to be the way it is. At some point around 40 million years ago, a mutation in an early lemur would have resulted in a very slightly elongated middle finger. It is very unlikely that this was a macromutation – a freak baby born with a noticeably extended central digit. Indeed, most evolutionary biologists today believe that such mutations will have played little role in evolution because they will almost certainly be detrimental to survival. It was probably an almost unnoticeable change, and may have conferred little if any advantage at first, but the possessor of the mutation certainly survived long enough to pass it on to its offspring. This is a very important aspect of evolution by natural selection. Mutations do not have to be immediately advantageous to survive in the population. Given time, however, those aye-ayes with slightly longer fingers must have enjoyed a slight advantage over their less well-endowed brothers and sisters. Perhaps they could spend slightly more time high up in the canopy because they could access easily accessible bugs in cracks in the wood. Thus, this particular subset of the genetic database carried in the proto-aye-aye population, with the lengthened middle finger mutation, began to separate from that of the other lemurs, who spent less time on the tree

270

trunks foraging for bugs. Over long periods of time, separation, further mutations, and continued selection pressures conspired to sieve this database in a different way to that of the lemurs without the finger mutation. And, eventually, the ancestors of the aye-ayes become so separated from the other lemurs that they are today clearly identifiable as a separate species. The forest canopy therefore acts in the same way as the physical island of Madagascar. It allows for the separation of a subset of the genetic database of life, which allows that subset to diverge from the rest in response to mutations and selection pressures. Because the organisms themselves are the indirect physical expression of the information in the database, these too will diverge in form and behaviour from their cousins on the isolated forest floor. Eventually, given sufficient time, isolation and perhaps selection pressure, this leads to the origin of new species. Species are not the direct product of natural selection, they are accidents – a by-product of the myriad genetic changes that have accumulated through natural selection and random genetic effects.

As an aside, it is somewhat difficult to come up with a precise definition of what constitutes a species. We all have a very rough understanding – a lion is a different species from a human being. A species is generally defined as a population of inter-breeding organisms that produce fertile offspring. However, species are not fixed, and in some cases they can even exchange genes, through the appearance of hybrids (although these are generally sterile, especially in animals) or through the transfer of genes between quite disparate organisms (this can occur in prokaryotes, but is so rare as to be biologically insignificant in multicellular organisms). The landscape of the global genetic database is constantly shifting in response to mutations, changing selection pressures and so on. Species become extinct, new niches open, populations become isolated for all sorts of reasons, and over time new species emerge.

As evening approached, we settled our little family unit of aye-ayes into their cage and walked back into the jungle. Aye-ayes build nests, and the team had carefully noted the location of their home. We opened the cage at the base of the tree and watched as first the mother and then the child climbed away into the darkness. These animals, clumsy and weird in our world, are elegant and graceful in their own, as they must be because their form was shaped by that world. Their clade, a single branch of the tree of life, has been separate and distinct for 40 million years. I feel this allows me to call the aye-aye 'ancient', even though I know that it is just as modern and evolved as you or me. So, this ancient animal, whose little piece of the great genetic database of life carries the tale of the traveller's raft and the imprint of a long history in the tangled Madagascan forests, vanished into the night, taking with it a unique story that, if lost, will never be replaced. ◉

**LEFT:** The aye-aye is supremely adapted to life in the forest canopy. Its suitability for this type of habitat developed over a very long period of time, through gradual mutations and continued selection.

# PRECIOUS ISLANDS

The last day of the ten months spent filming *Wonders of Life* was passed on a tiny island close to an island off the coast of an island country in a blue ocean on a unique and precious island. Our base was an old lighthouse, perched on a rocky outcrop in the Mozambique Channel, no bigger than a football pitch. This little world, with room for only five of our film and production team, is perched a five-minute boat ride from Nosy Be, itself a small one-town island off northern Madagascar. Islands are a powerful, multi-layered metaphor for the ideas presented in *Wonders of Life*. As we have seen, islands, which may be geological or simply isolated parts of the ecosystem, allow for the separation of gene pools – the hiving off of fully functional but incomplete parts of the great database of life – which will then change independently, leading to the emergence of new species. For Darwin's bark spider, the air space above a river is an island. For the *Termitomyces* fungus, the air-conditioned termite cities are islands, and in turn the fungus forms part of the island that produced the species of termite that farms it. Islands serve to shape the organisms within them by acting as a sieve, selecting for and against mutations and changes in the genetic codes of those organisms. Islands, therefore, are more than just geography; each one provides a complex, abstract collection of filters; it will impose physical separation, force different organisms together, isolate parts of the genetic database and continually redistribute that ever-shifting, mutating collection of information among a group of animals, plants, bacteria, fungi and archaea unique to that island.

The group of organisms on the island is thus a part of the island, which serves only to add to the complexity and drive forward the emergence of diversity. This is an extremely important point. Biodiversity creates islands, which in turn create diversity. The opposite is also true. The loss of species and habitats reduces the number of islands available, and therefore reduces the ability of the entire genetic database of our planet to shift in response to the constant storm of environmental challenge and change delivered by a living and geologically active planet in an unstable orbit around a

*Given there are over half a trillion galaxies in the observable Universe, the idea that there are no other planets out there with webs of life at least as complex as our own seems to me to be an absurd proposition.*

273

variable star, populated by a fizzing mix of competitive organisms that have no conscious concern for the maintenance of the great library of life. Except for us. We alone have evolved to understand how our island came to be alive, and the importance of the maintenance of its diversity. If this sounds overly sentimental or preachy, I do not intend that. I'm no part-time eco-warrior adorned with ethnic bracelets purchased from a street market in Glastonbury while on a gap year before embarking on a career in marketing or politics. I'm not concerned about protecting polar bears because they are white and fluffy, or a particular species of butterfly because it has uniquely patterned wings. In understanding a little more about evolution, however, I have become aware of the intimate connection between biodiversity and the capacity of the entire ecosystem to respond to unavoidable and constant challenge and change. It is self-evident that a reduction in the number of islands results in a reduction in the number of extant species today, and a reduction in the capacity of the ecosystem to produce new adaptations tomorrow. To me, this looks like a feedback loop that we should strive to avoid.

There is one more lesson I learnt from my year-long crash-course in evolutionary biology. I have come away with a strong sense that the basic biochemistry of life may have been virtually inevitable. Given the right conditions, the laws of physics not only allow living things to appear spontaneously, but may well make the emergence of life more likely than not on planets with a ready supply of liquid water in the presence of the thermodynamic driving forces of temperature and chemical gradients. And given that there are almost half a trillion stars in our galaxy alone, and over half a trillion galaxies in the observable Universe, the idea that there are no other planets out there with webs of life at least as complex as our own seems to me to be an absurd proposition. But that does not remove one iota of value from our web, because it is absolutely unique. Its development has been chaotic in both the colloquial and the technical sense. Tiny changes and chance happenings – from the endosymbiotic origins of the eukaryotic cell, to the crossing of the Mozambique Channel by a group of castaway lemur ancestors – have had dramatic impacts on the nature of Earth's great genetic database and its physical expression in the endless forms most beautiful that have lived, are alive today, and will live in the future. Every square centimetre of our island is definitely unique, and therefore uniquely valuable. Go outside, now, and look at any randomly selected piece of your world. It could be a scruffy corner of your garden, or even a clump of grass forcing its way through a concrete pavement. It is unique. Encoded deep in the biology of every cell in every blade of grass, in every insect's wing, in every bacterium cell, is the history of the third planet from the Sun in a Solar System making its way lethargically around a galaxy called the Milky Way. Its shape, form, function, colour, smell, taste, molecular structure, arrangement of atoms, sequence of bases, and possibilities for the future are all absolutely unique. There is nowhere else in the observable Universe where you will see precisely that little clump of emergent, living complexity. It is wonderful. And the reason that thought occurred to me is not because some guru told me that the world is wonderful. It is because Darwin, and generations of scientists before and after, have shown it to be so. ◉

*Look at any randomly selected piece of your world. Encoded deep in the biology of every cell in every blade of grass, in every insect's wing, in every bacterium cell, is the history of the third planet from the Sun in a Solar System making its way lethargically around a galaxy called the Milky Way. Its shape, form, function, colour, smell, taste, molecular structure, arrangement of atoms, sequence of bases, and possibilities for the future are all absolutely unique. There is nowhere else in the observable Universe where you will see precisely that little clump of emergent, living complexity. It is wonderful.*

**RIGHT:** Our planet's present is inextricably connected to its deep past through the mechanisms of evolutionary biology. The prognosis for its future – and for ours as human beings – is truly exciting.

# INDEX

Page numbers in *italics* indicate photographs and images

**277**

# K

# L

# M

283

# T

# U

285

# PICTURE CREDITS

Top=t; middle=m; bottom=b; left=l; right=r

# ACKNOWLEDGEMENTS

*Wonders of Life* was an unusual project for me in that I had to learn much of the material for the first time. This was a supremely enjoyable experience, not least because I had two excellent teachers – Dr Nick Lane from University College London and Professor Matthew Cobb from my home institute, The University of Manchester. I first met Nick at the Royal Society in 2010, when his book *Life Ascending* rightly beat my book, *Why Does e=mc2?*, to the Royal Society prize for Science Books. I was introduced to Matt by my ever-understanding vice-chancellor, Professor Dame Nancy Rothwell, who is unswervingly supportive of academics who want to devote a proportion of their professional time to the popularization of science. I thank all three unreservedly. I also thank The Royal Society, who have been equally supportive during my time with the BBC.

The television series would of course not have been made without an incredibly talented and dedicated team. We'd like to thank them all for the passion and thought that they crafted into *Wonders of Life*. We'd especially like to thank James van der Pool, the Series Producer, for his calm and intelligent nurturing of the series. Michael Lachman, Stephen Cooter, Paul Olding and Gideon Bradshaw for the world class directing and producing that they brought to the films. Kevin White and also Tom Heywood for their outstanding photography. George McMillan and Christopher Youle-Grayling not just for sound but for everything else that a great sound man brings to a team and the production managers Jenny Scott and Alexandra Nicholson for the skill and patience they brought to managing such a complex production.

There are of course so many others who helped make the series but we would also like to thank Ben Wilson, Suzy Boyles, William Ellerby, Rebecca Edwards, Helene Ganichaud, Weini Craughan, Leili Farzaneh, Graeme Dawson, Simon Sykes, Martin Johnson, Gerard Evans, Darren Jonusas, Louise Salkow, Laura Davey, Laetitia Ducom, Matt Grimwood, Shibbir Ahmed, David Schweitzer, David Maitland, Michael Pitts, Neil Kent and Nicola Kingham.

I would also like to make special mention of the BBC, the only broadcaster in the world who can produce a series like *Wonders of Life*. Public service broadcasting seems to be under constant attack from certain quarters. Some of the shrill voices can be identified with vested interests, and these have my utmost contempt. Others genuinely believe that 'choice', by which they mean an ever expanding multi-channel environment, with viewers selecting precisely what they want to watch and paying appropriately, is the future of broadcasting. This, in my view, is a recipe for ghettoization and as such would do a profound dis-service to the viewer. I discovered, by chance, that biology is a fascinating subject, and understanding it has enriched my life. This is the fundamental role of a publicly funded, public service broadcaster. It is Reithian to the core. It is enriching for viewers to chance upon programming they wouldn't necessarily choose to watch. This is the foundation of education, and I see no reason why television should not be considered, at least partly, as a pillar of our public education system. Genuine choice comes with breadth of experience, not with the provision of 24-hour sports channels. I hope at least some viewers will stumble upon *Wonders of Life* and discover accidently, as I did, that learning about the life sciences is a tremendously entertaining and valuable way to spend a Sunday evening.

Finally, and most importantly, I thank my family Gia, Mo and George, who did not choose to share me with *Wonders* for the last four years. I promise to take a break now!